CITIES

ESSAY SERIES 50

Guernica Editions Inc. acknowledges the support of The Canada Council for the Arts.
Guernica Editions Inc. acknowledges the support of the Ontario Arts Council.
Guernica Editions Inc. acknowledges the financial support of the Government of Canada through the Book Publishing Industry Development Program (BPIDP).

EDMUND P. FOWLER

CITIES, CULTURE AND GRANITE

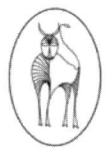

GUERNICA
TORONTO – BUFFALO – CHICAGO – LANCASTER (U.K.)
2004

Copyright © 2004, by Edmund Fowler and Guernica Editions Inc.
All rights reserved. The use of any part of this publication, reproduced, transmitted in any form or by any means, electronic, mechanical, photocopying, recording or otherwise stored in a retrieval system, without the prior consent of the publisher is an infringement of the copyright law.

Antonio D'Alfonso, editor
Guernica Editions Inc.
P.O. Box 117, Station P, Toronto (ON), Canada M5S 2S6
2250 Military Road, Tonawanda, N.Y. 14150-6000 U.S.A.

Distributors:
University of Toronto Press Distribution,
5201 Dufferin Street, Toronto, (ON), Canada M3H 5T8
Gazelle Book Services, White Cross Mills, High Town, Lancaster LA1 1XS U.K.
Independent Publishers Group,
814 N. Franklin Street, Chicago, Il. 60610 U.S.A.

Typeset by Selina.
Printed in Canada.
First edition.

Legal Deposit — Second Quarter
National Library of Canada
Library of Congress Catalog Card Number: 2004101873
National Library of Canada Cataloguing in Publication
Fowler, Edmund P. (Edmund Prince)
Cities, culture and granite / Edmund P. Fowler.
(Essay series ; 50)
ISBN 1-55071-194-6
1. Urbanization. 2. Urban ecology. 3. Urban policy.
I. Title. II. Series: Essay series (Toronto, Ont.) ; 50.
HT122.F69 2004   307.76   C2004-901102-2

Contents

Preface 7
Foreword by P.A. Dutil 9
The Ecological Sense of Mixed Land Use 13
Reflections on *Building Cities That Work* 35
The Politics and Ecology of a Healthy City 47
Is It Accidental or Merely Wild? 59
Getting Urban Growth Wrong 71
Domestication as Pavement 86
What is Sustainability? Heidigger in Hamilton 102
Culture and Granite 111

TO JANE JACOBS

Preface

This book is dedicated to Jane Jacobs because her ideas have stimulated so many of my own. They lay at the heart of my book *Building Cities That Work* and figure prominently in many of the essays in this volume. She has advised and encouraged me so many times in my work that it is difficult to overstate my debt to her.

Another source of help and provocative ideas has been John Livingston, whose profound insights have transformed the way I regard nature and the part of it that is human.

Many of these pieces are book review essays that Pat Dutil invited me to write for the *Literary Review of Canada*. I am enormously grateful to him for inviting me to explore writing about social science in a more literary style. Indeed, without this invitation, this book would never have seen the light of day. On the other hand, the mix of styles did not interest most publishers, who could not place it in a convenient category. I am fortunate to have found Antonio D'Alfonso, whose publishing programme at Guernica Editions celebrates this kind of cross-fertilization.

I also wish to thank Pat Dutil, Bob Gibson of *Alternatives*, and Richard McCarthy of *Blueprint for Social Justice* for editing skills that improved the logic and flow of the articles that appeared in their journals.

My wife Shelly's ideas and intelligence are woven into every essay of this book. She is a true inspiration.

The author and publisher wish to express their thanks to the following publications for permission to reprint articles, or earlier versions of these articles, from their pages. *Alternatives Journal: Environmental Thought, Policy and Action*, Faculty of Environmental Studies, University of Waterloo, Waterloo Ontario N2L 3G1: "Land Use in the Ecologically Sensible City," Summer, 1991. *Blueprint for Social Justice*, Twomey Center for Peace Through Justice, Loyola University, Box 12, 6363 Charles Ave., New Orleans, LA 70118-6195 : "Reflections on Building Cities That Work," April, 1994; "The Politics and Ecology of a Healthy City," April, 1996. *Literary Review of Canada*, 581 Markham St., Suite 3A, Toronto, Ont. M6G 2L7: "Pavement Mentalities," November, 1995; "The Making of Toronto," September, 1996; "Getting Urban Growth Wrong," March, 1999; "Heidigger in Hamilton," March, 2002.

Foreword

It is a truism that big cities – notwithstanding their determining impact on a nation's prosperity, cultural life and intellectual vitality – exercise remarkably little political power. Almost without exception, central political systems have systematically worked to diminish their voices. Great cities such as Paris and London have only elected their own mayors for the first time within the last twenty years. In Canada, cities are creatures of provincial governments. While denizens have long enjoyed the right to elect their own municipal councils, mayors and city councilors must work within the laws, guidelines and regulations imposed by the provincial and federal government.

Perhaps for this reason of powerlessness, little is expected from them. Municipal policy rarely gets front-page coverage in newspapers and local politics cannot compete with the myriad tragedies that are standard fare for dinnertime news programs. It is not surprising to note, in this light, that municipal electoral contests fail to win the attention of voters. In Canada, the turnout for local elections is embarrassingly low. People are simply convinced that they have no impact on the governance and administration of the city they live in. The result seems preordained: cities struggle from crisis to crisis, impaired by a policy-formulation apparatus that cannot cope with the greater realities it must deal with. Members of council, shorn of the necessary policy tools needed to deal seriously with larger issues, remain focused on pothole issues that relate directly to the short-term needs of the few constituents that will vote for them.

In this loveless affair, thinkers have often played a critical role, and their teachings, with time, have influenced occasional waves of reformist politicians and administrators. But with every success, cities have also shown that they cannot independently cope with some of the greatest issues of day, namely environmental impacts and sustainability.

Waiting for the next generation of progressives, Edmund Fowler ranks among the leading Canadians who are speaking out for a renaissance of the city. He rails against all that is plastic – from plastic language to plastic cities, indeed to the very production of plastics destined to fill landfills. His plea for an enlightened community that comprehensively addresses its environmental impact must be read and acted upon.

Edmund Fowler loves cities. He enjoys their economic vigour, their promise of community and their cultural creativity. In 1992, he published *Building Cities That Work*, and his impassioned plea for better long-range planning impressed me with its vitality. At the time, I was the editor of the *Literary Review of Canada*, a monthly magazine I had founded the year before. Fowler has remained a steadfast and generous contributor to the magazine over the past decade, and many of the articles he first published in the *LRC* are now bound within these pages.

I launched the *LRC* to give Canada's thinkers a consistent and reliable vehicle to exchange ideas. It was the first magazine of its kind to give authors an opportunity to discuss the views and hypotheses presented in books through exhaustive book reviews.

These were troubled times in Canada. Two major constitutional initiatives failed; a major recession stalled commercial initiative; federal, provincial and municipal governments embarked on draconian cutbacks to maintain a budgetary balance. In 1993, Canadians elected the weirdest parliament ever in Ottawa: the avowed separatist party Bloc Québécois was now the official opposition, beating by a nose the Reform Party, a party based exclusively in Western Canada. The NDP and two Tories brought up the rear as irrelevant versions of the important parties they once were. Cities in this context were forgotten.

Many thinkers responded to my call for enlightened discussion of books and issues, and I'm proud to have counted Fowler among them. His essays reveal the insights of a thinker not afraid to dream of the possible. His objective is nothing less than cities that are clean, vibrant and attractive to a wide variety of people from all walks of life. These are cities that work from below, where members of the public are confident that their voices will be heard and acted upon. At one point, he calls cities "concrete expressions of our dominant values" and he wants to ensure that those values are the correct ones. He wants planning to take place in the broadest context possible – one that considers the environment as vital not just because cities do the most to alter it, but also because cities will be stronger if they genuinely contribute to diminishing their impacts. He also wants livable cities: places with safe, enticing street lives that will attract business people, workers, thinkers, and artists.

In rereading Fowler, however, I am struck by his battle against time. He draws from history to capture the best lessons from past practices; he urgently points to policies from the four corners of the earth to prove what is possible; he is anxious for the race that seems to be lost in preserving the environment. He enjoins a battle for the minds of the city and patiently pleads for an urban planning that is enlightened in its use of space and place. He wants what we all want but find difficult to articulate: a sensible city where one can breathe the air, drink the water, live in peace. But Fowler does not stop there. Taking seriously these ideals, he fearlessly prescribes solutions that are bold and imaginative.

I invite readers to consider his arguments for a new city and to be challenged by Fowler's clarion call to treat the city as a "whole" organism. I suspect that years from now, historians will turn to our time to examine the life and death of cities and will consider Fowler's work. Inevitably, they will ask: Was his work of consequence? Only you, the reader who holds this book in your hands, will be able to answer that question. My hope is that you will contribute to making the thoughts contained in this book influential by applying them, referring to them, and making them "politic." As cities are increasingly recognized for their economic, environmental, cultural and social impacts it is incumbent on us to ensure that our politics make them healthier.

P. A. Dutil
Toronto, Ontario

The Ecological Sense of Mixed Land Use

Land use planning, like architecture, is considered the preserve of experts and of impersonal market forces, both of which, in turn, are thought to be properly independent of the so-called ordinary citizen.

It is seldom acknowledged that the process by which cities regulate land and buildings is a subjective and detailed reflection of unexamined assumptions, confusions, traditions, ideals, myths, and predilections about human nature and social behaviour. Instead, we relegate the conscious part of the land regulation process to a subgroup of municipal planners and overworked city councilors. These people who are on the front lines, so to speak, are far too busy processing applications for re-zonings and big developments to pay attention to larger questions.

This is no longer acceptable. These larger questions – about how to avoid ruining the planet's ecosystem with our development practices, how to reverse the alarming trends of global warming, destruction of the ozone layer, and dramatic increases in the pollution of air, water, and soil – have become urgent.

Land use decisions by local governments are daily adding to environmental damage, largely because most decision-makers are unaware of the many interconnections between urban land use regulation and the biosphere. At least in North America, the central role of local governments is the protection, servicing, and regulation of property. When local officials talk about what they are doing in the way of environmental policy, they point to recycling programs, tree planting, and banning chlorofluorocarbons. And while such policies are significant, they ignore the basic patterns of land use that must be changed before our cities can be ecologically sensible settlements.

Being unaware of our connections to the built environment is part of the same blindness that makes us unaware of the "natural" environment. In both cases, the blindness has played a role in the increasingly dangerous environmental crisis we confront.

There is an urgent need for land use decision-makers to recognize the not-so-obvious links between municipal land use and deterioration of the environment. In fact, I wrote this article originally to present to land use committees of local governments – to suggest what they could do, now, in the way of positive action, with the applications they are considering on a daily basis. In that spirit, very specific proposals are offered, proposals whose force lies more in their concreteness than in their comprehensiveness, more in their technical practicality than in their attractiveness to local politicians (whose adherence to environmental causes varies widely in any case).

There will obviously be political resistance to these proposals. In fact, when I spoke to the different committees, reactions were predictably negative and even uncomprehending. This resistance is to be expected, because many of us are inured to the postwar urban environment and we find it hard to imagine anything else. It has a kind of psychological and political hold over us. This means that if we are to achieve the ecologically sensible city we shall have become aware of how our attitudes and habits of behaviour towards each other and towards the environment are influenced by that environment. We shall have to recognize the psychological, social, and political dimensions of ecologically unsound cities.

The Pattern of Urban Land Use

Since World War II, we have been building cities in North America in a distinct way. The design has been copied worldwide, but here we have developed it more massively than elsewhere. Basically, we have been building deconcentrated cities, with segregated land use and very large scale developments.

It is no secret that our cities have spilled out over the countryside and now take up hundreds, sometimes thousands of square kilometres. Geographers have demonstrated empirically that the density gradient of North American cities – how quickly the population thins out as one travels from the centre – has become extremely gradual. And the intensity of activity per hectare in our cities is far lower than in other parts of the world.

The concept of zoning urban land for different uses has been around at least since the 19th century, but we have built cities and suburbs where land uses are *completely* separated. With few exceptions (such as some high rise developments in suburbs that cluster around shopping centres) one has to travel considerable distances from one's dwelling in order to shop. The number of people who can walk to work from where they live is very small. Indeed, as I shall argue below, we no longer expect such a thing.

In addition, urban development has become almost unbelievably large scale. Residential communities that house over 50,000 people, built by single development companies, are not uncommon. Neither are shopping centres with over 200 stores, factories that cover hundreds of hectares, or multi-building office projects thirty, forty, and fifty stories tall.

Urban Land Use and the Environment

These characteristics of the postwar North American city are closely connected to our current environmental headaches. The car is the clearest example: segregated and deconcentrated land use was developed in a symbiotic relation with motorized vehicles. Which came first at this point is immaterial. What is important is that we have sunk vast amounts of capital into a built environment that requires us to use cars and other means of medium-range transportation simply to be able to function.

The air pollution that is killing humans (as many as 50,000 a year in the United States) as well as lakes and forests – the air pollution that is producing a greenhouse effect with potentially catastrophic effects on coastal populations and economies – this air pollution is produced to a large extent by motor vehicles in cities. We must, of course, continue to put restrictions on car emissions, increase their fuel efficiency, and coax people to use public transit and bicycles, but unless we rethink totally the land use policies that make car use so necessary in the first place, the other effects will have little impact.

Although it is becoming more common to acknowledge that transportation and land use policies are connected, few policy makers seem to have grasped how closely they are intertwined. In most large North American cities, for example, transportation and land use planning are carried out in completely different offices by different sets of people. Nevertheless, every time a land use committee requires parking spaces for a development, or approves a subdivision whose density and location make the car the only feasible means of transportation, or even suggests a modification to decrease traffic congestion around a development, they are encouraging automobile use and contributing to the air pollution that blankets our cities. Thus, one clear principle of land use in the ecologically sensible city is to discourage the use of the private car.

While the car is an obvious means of transporting human bodies around our deconcentrated and segregated cities, other items must be transported as well. Energy is one example. First, our single use areas are all used at different times. Vast areas of the city remain largely unused during weekdays, and only come to life at night, or on weekends – and vice versa. We are, in effect, heating in winter and cooling in summer two separate built environments, because of large, single-use developments.

Second, we "transport" energy to our homes and workplaces. It is produced in immense power plants, far from where it will finally be used. Aside from the environmental hazards of these large installations themselves, distribution requires a vast network, with millions of kilometres of wire, thousands of kilometres of rights of way, transformers, repair personnel and their trucks, transportation of fuel and so on. The fact is that the energy industry *itself* consumes 25% of all the energy produced in North America. (Transportation consumes 20%.) Thus, using some land just for energy production and then delivering it to spread out cities results in enormous air pollution (from coal-fired generators), unimaginable risks from nuclear waste, destruction of thousands of square kilometres of virgin wilderness in the case of hydro-electric power, and wasteful use of energy itself. The obvious land use policy is to require local developments to be energy self-sufficient. The technology for this is available; we simply need the will to make it obligatory in all new buildings. This would be another principle of the ecologically sensible city's land use policy.

The use of land for agriculture in North America also reflects the pattern of land use segregation and deconcentration. We have been convinced that our food must be grown in large, separated agricultural "factories," even though reliable estimates show that the average city dweller is capable of growing a sizeable proportion of his or her own food within the city.

Hong Kong, one of the world's most densely populated settlements, grows 45% of its own vegetables. Here we rely on a huge production and distribution system that is ruining the earth's topsoil, poisoning its rivers and lakes, and covering the urban landscape with larger and larger supermarkets. Ironically, as Wendell Berry has remarked, the huge farms are no more self-sufficient than the cities that depend on them. The farms need city-produced machines and fertilizers, and the farmers go to the supermarket to buy food.

It would obviously make more sense to set aside land in cities for growing food. In warmer climates, agriculture can take place all year round. Advances in permaculture and greenhouse design are now making it possible to do this in northern areas as well. The problem has been that growing a head of lettuce in a conventional greenhouse in Ontario uses ten times the energy needed to grow it in California and ship it to eastern Canada. John Todd of the New Alchemists has come up with a way to produce vegetables in greenhouses throughout the winter in a cold climate, mainly by incorporating fish tanks, which also provide fertilizer:

> Our larger greenhouse has been operating for two winters; it maintains temperatures well above freezing, even on nights when the outside temperature drops to -20 or -25 C. The only energy which must be expended is electricity for a circulating fan, and no supplemental heating has been required ... Over 60 percent of the winter heat storage comes from warmth stored in the large translucent fish tanks ... The greenhouse solar design has enabled us to produce winter crops of lettuce, broccoli, chard, and other cool weather plants – heating only with the sun. This year we harvested tomatoes in spring and early summer with yields roughly equivalent to those *in conventionally-heated greenhouses*. (McRobie, 173)

Cooperative greenhouses have already been successfully developed in northern urban milieux as diverse as Cheyenne, Wyoming and Chicago, Illinois. Dolores Hayden has also shown how a city's backyards, alleys, and service entrances can be reorganized to be developed as cooperatively-owned open space for gardeners. Thus, municipal regulation of land use could involve encouraging small-scale agriculture on rooftops and undeveloped open space.

Big city waste problems are also a function of sprawl and of land use segregation. As our cities became bigger, we usually just built bigger waste disposal systems. The increasing amounts of garbage simply had to be trucked further and further away. Just as with energy, we are expecting not to have to deal with the garbage problem. We ship it elsewhere. Our land use reflects this attitude, for our "individual" dwelling units and workplaces in general make no provision for disposing of their own waste, although this is beginning to change.

Patterns of urban deconcentration made the use of waste disposal sites on the fringes of the metropolis a plausible solution to the problem of getting rid of the city's garbage. The habit of segregating land use fits right into the policy of segregating garbage dumps. And large-scale developments, especially high rises, mean large scale garbage: on-site disposal seemed out of the question. Yet in Japan, in certain prefectures, household waste is not picked up at the front door (except for paper and fabric). Everyone takes his or her garbage to the local recycling centre about a block away, already separated into compostable kitchen waste, plastics and incombustibles, glass, and cans. The physical possibility is obviously there. Land and space in buildings must be reserved for the garbage we produce. Small scale recycling plants can be just as efficient as large ones.

Although North America's municipalities tend to be responsible about treating their sewage, our water supply is still being poisoned – by municipalities that are not so responsible, by factories, and by other land uses that regularly dump noxious effluents into what would otherwise be clean water. Because so much city land is paved, and because the pavement collects animal excrement, dirty oil and rubber from cars, and other nasty things that people throw on the street, storm sewers deliver dirty loads of water into lakes and rivers every time there is a heavy rainfall. More concentrated and mixed land use would not only make it less likely that irresponsible acts of pollution would occur, it would also decrease the amount of paved land.

As well, new projects can be required by land-use committees to install their own sewage treatment. This is not as far-fetched as it sounds. John Todd has demonstrated the workability of greenhouse-based sewage-purifying ecosystems that look and smell like a botanical garden. Todd has designed a system for the town of Harwich on Cape Cod that took septic wastes, restaurant fats and greases, hospital wastes, and industrial effluents, which included fourteen out of fifteen of the most dangerous toxic volatile organic chemicals. Using micro-organisms and plants such as water hyacinths, Todd eliminated 100% of thirteen of these pollutants (including hydrocarbons, dioxin, and PCBs) and 99.9% of the fourteenth. E. coli bacteria were one one hundredth of those allowed in swimming water.

In using this technology, we need not think of large-scale systems. Todd built one at a ski resort in Vermont, with a capacity of 57,000 litres, which would fit neatly into a small urban space without taking up more than 40 to 50 square metres.

Significantly, Todd has designed ways of incorporating such systems right into city streetscapes. Treating sewage and other toxic effluents close to their sources could easily become part of the land use policy of planners and councilors seeking to make their cities more friendly to the environment.

Thus, while deconcentrated, segregated, and large-scale land uses are part of the pattern leading us towards an environmental crisis, the technical solutions of small-scale mixed land use with a concentration of activity, combined with on-site energy installations and garbage and waste disposal are all easily attainable. But it hasn't happened. This is because the psychology of implementing such solutions is a larger obstacle than any technical one. This obstacle is, significantly, also related to the design of the physical environment.

The Impact of the Built Environment

Our built environment has to be understood as an essential part of the psychology of our wasteful consumption habits. For the most part we do not comprehend the significance of this. Instead, we take as given the pattern of city streets, the distribution of land use, and the design of buildings and factories. However, our attitudes and behaviour are shaped to an astonishing extent by these patterns and designs. Workers who live 20 kilometres from their jobs (and there are many) will laugh at anyone who tells them it would be better environmentally if they rode a bicycle to work. Most residents of high rises would find it difficult to reduce air conditioning costs by planting shade trees. A good example of how we are conditioned is provided by Ruben Nelson:

> Imagine that you work downtown in a large Canadian city . . . How often are you likely to get home for lunch? Not very often, right? If you work downtown in a large city, it is almost impossible to get home for lunch. Most of the time you won't even try. You won't formulate the intention of getting home for lunch because of what you know about life in the city. The structure of our large cities defines not only whether we actually get home for lunch, but whether it will even occur to us to try to do so. If we live in such a city long enough, neither we nor our families form the expectation that we will be home for lunch . . . [W]e learn in time, if we live with the arrangement long enough, we will likely forget what we are missing and see the present arrangements as "normal." (33)

Consider how this logic applies to the connection between urban design and our attitudes towards the environment, attitudes which – at least up to now – have been considered perfectly normal. By the way we segregate our land, we put waste disposal out of sight and therefore out of mind. If we had it on the corner, like the Japanese, we would think more carefully about our garbage-production habits. This is especially true of toxic waste. Seeing piles of poisonous substances down the block is far more consciousness-raising than being able to hide them in a plastic bag for pickup by an anonymous truck. Thus, the way we use land for garbage disposal is linked to our problems with too much waste and nowhere to put it.

Just as is the case with food (discussed above), we use countless products whose manufacturing process is unknown to us, because the factories are separated from us by dozens and even thousands of kilometres; the toxic chemicals used in making toilet paper and plastics are not evident to us because of this separation. Nevertheless, because everything is connected on this planet, after many years of increasing production of these substances, the toxic runoff and leachate is now poisoning the water we consume in cities. Every year, three million tonnes of mercury and six million tonnes of oil are poured into the world's rivers.

If this were the only toxic waste problem, it would be serious enough. But our consumption habits have produced a much worse monster: hidden toxic chemical dumps such as the Love Canal in Niagara Falls, New York. There are thousands of such dumps across North America seeping poisons into aquifers, rivers, and lakes. While in a sense the damage has already been done, we have to see the lesson, that only a large-scale plant, far away from other land uses, could be capable of such environmental atrocities. Deconcentration, separation of use, and large-scale development are related not just to the production of toxic waste but to our awareness of toxic waste.

Segregation of large-scale, sophisticated factories helps to hide other unpleasant things from our senses. Electricity is touted as clean energy, but although an all-electric house looks squeaky clean, somewhere miles away are power plants providing that energy, all the while leaking their deadly wastes to the air, water, and soil.

Computers, too, are deceptive, since their manufacture involves notoriously toxic chemical baths, with the results that factory workers suffer disproportionately from a variety of serious cancers, and effluents from factories pollute groundwater and make it undrinkable.

In general, big, distant manufacturers use tremendous amounts of water to make the products we buy. Many of us know now that the average home toilet uses 19 litres per flush, and that we could cut that down substantially. Perhaps fewer of us know that most households use 379 litres of water a day in Canada. But if we add up all water used *per person* in Canada, it turns out to be 4,129 litres each day. (In the United States, 6,318 litres are used per person per day, which ranks that country number one in the world in water use; Canada is number two.) Where does all this water go?

It takes 303 litres of water to produce *one* soft drink can. It takes 683 to produce *one* newspaper. And if we come full circle back to the automobile, it takes 454,560 litres to produce a single car. All this water comes in clean, of course, and goes out polluted in one form or another. But because of the way we have separated the production process from where we live and work, it is hard to be aware of how all this water is used.

There are other ways the built environment conditions us so that our behaviour harms the environment. North America's deconcentrated, segregated city has encouraged us to want mobility – the ability to get *to* what we need – rather than access, which is having what we need where we need it. Streets are treated as conduits for the passage of motorized vehicles from one part of the city to another. Most of us think of them this way, as paths, rather than as places where things happen. There are exceptions, of course: many cities have pedestrian-oriented strips and neighbourhoods. But the more we build streets that are conduits (or "arteries" or "feeder" streets), and the more we conceive of the city as an urban agglomeration to get around, the less importance we give in our everyday life to streets and sidewalks as things in themselves, as places where important events happen. Neither we nor many of our planners think of them primarily as places where we shop, visit, observe, or possibly find work. Most of the building going on now, which is in the suburbs, is creating dead streets with no street life. This is not just bad planning in itself, it is ominous because more and more of us think of such streets as "normal," as Ruben Nelson incisively argues.

The irony is that, as tourists, we flock to places such as Quebec City and Savannah, Georgia, or even to Toronto's Yorkville, because they offer the intrinsic appeal of narrow streets and pedestrian-oriented businesses. Building cities like this would be far less damaging to the environment because they do not need cars and because they use municipal services such as utilities more sparingly. Such places would, considering their popularity, be much more fun to live in, but we simply do not see it as a choice. It is not easy to notice how the built environment conditions us to think of only a few alternate urban forms, which are those we have grown up with. If those urban forms were environmentally sensible, it would just be a case of harmless delusion, but they are not.

Being unaware of our link to the natural environment is thus mirrored in our insensitivity to our own built environment. What we have built only reinforces and perpetuates our feelings of separateness from our urban surroundings – streets, sidewalks, parking lots, huge buildings, and garish commercial strips. Planning the ecologically sensible city will involve being aware of the interpenetration, as Nelson calls it, of environmental design and our psyches.

The direct impact of the design of the postwar city on human behaviour has been demonstrated in my own study of nineteen different Toronto neighbourhoods. These neighbourhoods ranged from the physically diverse – with mixtures of land use and of old and new buildings, medium to high concentration of use, and short blocks – to physically homogeneous areas, such as those with only residential high rises, or only suburban housing, or only warehousing. The less overall small-scale physical diversity, no matter what the socio-economic makeup of the area, the less neighbours knew each other, and the more crime, especially juvenile crime, there was. It is important to note that this relationship held up in the suburbs as well.

In some analogous findings, Oscar Newman has shown how large-scale high rises have many more problems with crime than other housing designs that give users ample opportunity for informal contact with each other in entranceways and other semi-public places. In general, the higher the high rise, the more crime in and around the buildings, especially vandalism, a crime directed *against a building*. Vandalism tends to be the work (or play!) of juveniles.

High rises are the perfect illustration of how easily one kind of urban design can wipe out street life, replacing sidewalks with miles of anonymous corridors and parking garages. The only way such places can be kept safe is with expensive and elaborate security systems, which are possible only in luxury buildings. Casual, civilized public life is ruled out.

Our built environment undercuts the daily experience of public life in cities. City and suburban land use patterns both reflect and reinforce the illusion of autonomy and withdrawal from the street. The suburbs were built for families with one (male) breadwinner whose spouse stays home and looks after two children, with a maximum of in-house conveniences and a minimum of informal neighbourhood street life. Dwelling units are built like pioneer cabins to be self-sufficient – at a price. Spread out, exclusively residential districts with single family dwellings require us to use the car to shop, and shopping turns into a major expedition to a big supermarket. Then we drive back home and dump everything into freezers and large refrigerators. In contrast, physically diverse neighbourhoods allow us, even encourage us, to shop daily for fresher though more perishable food on foot, because, by definition, smaller local stores are close. We can thus be more connected to the street, the local shopkeepers and neighbours. However, this is no longer considered a normal lifestyle.

In most of North America's suburbia, instead of sharing laundry facilities or playground equipment, each family has its own. Backyards are private enclaves for barbecuing and entertaining, and front yards are decorative defences against public intrusion rather than places from which to make a connection to the street. North Americans seldom think in terms of semi-public or public squares and courtyards where the civic life of a neighbourhood is likely to be nourished.

The postwar city, in short, helps to produce urbanites who are cut off from their immediate physical and social environments, but who have grown to think of it as normal.

The impact of urban design on our sensitivities to the environment is most clearly seen in children. Children are not as mobile as adults; they are often captives of their immediate physical surroundings. If they grow up in a neighbourhood with only manicured lawns, pre-designed playgrounds and buildings that get torn down instead of modified – in short, an unmanipulable environment – then they will learn that is the way things are and accept it as natural. If they feel unconnected to the environment, they are not going to take responsibility for it, just like their elders. If they do not have contact with nature, they will not develop an open and understanding relationship with it.

What children learn about relating to the built and natural environments is highly significant, for they are the adults of tomorrow, and those adults are even more likely to think of our artificial and unhealthy built environment as normal. Deconcentrated cities, with large-scale and segregated land uses, tell children, "This is the way things are built: they are built *for* you; don't mess with them, because they aren't supposed to be changed. You are just a consumer." As the research by Newman and myself shows, children react by being indifferent to their environments, or even committing more crimes and vandalism against it. As Bernard Rudovsky has suggested, in many cases the environment truly invites it.

Feelings of separation and powerlessness with respect to the environment are embedded in us, in part, by the design of our physical surroundings. The high rise, again, is a good example. How can we expect someone living in a high rise to start taking responsibility for the ecological consequences of her or his everyday life? The building already excludes it. One is prevented even from considering such choices as changing a window design, adding a room or balcony, growing a garden, or switching energy sources. Not impossible, but improbable, would be certain economic options: starting a day care cooperative, borrowing a vacuum cleaner from a neighbour, setting up a business or a workshop in the home. Socially, high rise dwellers are isolated from their neighbours. Psychologically, they are living in a fortress with dozens of mechanical devices to protect them from crimes which the building's design invites. Would it be any wonder if they also felt powerless? And yet millions of people *choose* to live in luxury versions of these concrete prisons.

Something inside of us must say it's OK. We must see that this syndrome of separation and powerlessness is somehow linked to the urban design typified by high rises.

The politics, ecology, sociology, and economics of it all fit together. If a city really works on any of these dimensions, it works on them all. To me, the implication is clear. If we as individuals do not take responsibility for the design and management of our own physical surroundings, policies to clean up the environment will be a farce – they will never mesh with our daily life experiences.

Doing It Ourselves

Consider the dozens of small scale self-help urban revitalization projects, whose energy came from ordinary people taking responsibility for their streets and their housing. At the low end of the socio-economic scale, the South Bronx has a number of local groups who are rebuilding their neighbourhoods. The South Bronx is one of the most desolate urban landscapes in North America, with hundreds of vacant lots, a population with many drug addicts and few financial resources, and very high crime rates. In the middle of all this, a small group of people started composting garbage, using the compost to fertilize the soil of some of the vacant lots, and growing their own vegetables. However, they did not stop there. Before long, they were renovating abandoned tenements with their own labour – "sweat equity" – and turning them into homes for themselves. One of these groups has asked for help in building a windmill generator. In another part of the South Bronx, another group started by fixing up a tenement and then turned to community gardening. (See McRobie and Gratz.)

The sequence of these local renaissances is not important. What is important are the numerous examples showing that when people get it together to start rebuilding their own neighbourhoods, they do it in environmentally friendly ways, growing their own vegetables, looking for alternative sources of energy, and renovating housing in a small scale way. Small scale buildings are environmentally sensible because they can always be retrofitted and redesigned – the city can change and grow without tearing down old buildings and using large amounts of new materials and experts to construct new buildings. The big high rises we have now are useless in this respect. They can never be modified to move with the times and are thus incredibly extravagant. They make no environmental sense. Land use committees who ignore this fact are simply being irresponsible.

One other dimension of sweat equity and community gardening in New York City must be noted. Although crime is a problem everywhere in that city, it is much less so where there are community gardens. Taking care of the soil and growing vegetables means caring more about each other.

User-designed communities called co-housing provide another illustration of people taking responsibility for their built environment. Popular first in Denmark, but now spreading to the rest of Europe and to North America, co-housing generally involves a group of five to twenty families deciding they would like to share a community, choosing a site (city, country, or the suburbs), designing the housing with the help of architects, sometimes helping with construction to keep costs down, and creating a small neighbourhood. These projects always have community buildings, where residents will often eat together, daycare programs are organized, television is watched communally, teenagers practise with their band, classes are taught, and all manner of other things go on. The housing units are modest, with few large appliances – everyone does his or her laundry at the community house. While privacy is always available in co-housing, the design ensures that there is plenty of opportunity for casual contact among neighbours in semi-public places.

Most significantly, most of these projects, apart from obvious consumption savings in the use of communal facilities, use some form of alternative energy. Environmentally sensible design is an important principle of co-housing.

It must be stressed that the organization and design process of co-housing is not all sweetness and light. It is a bit like organizing a wedding: even people who are desirous of co-operating beforehand discover they have all kinds of hidden predispositions and agendas that they were not even aware of. Anything that is user-designed is going to have these problems. It is part of the politics of the ecologically sensible city, a politics of learning how to work together, as opposed to a politics of large scale authorities, power struggles, and elections.

Ecologically Sensible Planning

The point of the above discussion is that planning the ecologically sensible city is not just a technical exercise. It involves social, psychological, and political transformations on the part of everyone involved (and perhaps even of some who are uninvolved). It cannot be imposed from above by enlightened authoritarians.

Following the argument of this article in somewhat reversed form, some principles of ecologically sane development could be the following:

- Let users design development;
- Favour small scale development;
- Mix land uses;
- Concentrate land uses;
- Seek self-sufficiency (food, energy, waste);
- Discourage the private car.

These principles – as well as the specific proposals listed at the end – are not, and cannot be, mutually exclusive or rigorous in a logical sense. They complement each other. That is, a city that mixes land use and concentrates it will have fewer cars, and fewer cars will lead towards mixed and concentrated land use. The two cannot be logically separated. Nor are these principles comprehensive. In fact, a claim to comprehensiveness would deny the principles' usefulness as characteristics of a city that fits into the biosphere. Such a city must have the same organic unity as a tree. Claiming I have a comprehensive list of principles for an ecological city would be like claiming I have discovered a list of principles for a tree.

The proposals sound incredibly authoritarian, because they are phrased in the kind of language we use for laws. We do not really need laws to tell us what comes to us naturally, but we have been building unnatural cities for so long that alternatives seem strange to us. As Christopher Alexander has pointed out in *The Timeless Way of Building*, we all have within us the ability to design buildings and cities that work, but we have forgotten how. Until we remember, there are some guidelines that can push us in the right direction. After that, they no longer need to be made explicit. (Think of the number of unwritten rules we obey!)

Actually, the idea behind the proposals is to move us away from authoritarianism in city building. Adopting them would push us to become involved actively in the physical design of our built environment and not to let it be built for us. City building has become simply another expression of domination of humans by other humans, which is exactly the relationship we adopt with respect to the "natural" environment. (I do not fully accept the distinction between the "built" and the "natural" environment.) Healing our relationships to the city becomes another way of healing our relationship to the planet.

The principles and proposals offered here are not a blueprint for the ecological city. Even if such a blueprint were desirable, we do not have the time to develop one in any detail. The situation is too urgent, and we must make use of the energy and money now committed to building and rebuilding in the city; otherwise we shall lose the chance to save ourselves, lost in a blizzard of planning reports and statements of vague goals. But the fact is that such a blueprint is not desirable. By definition, the ecological city is an organic process, without an endpoint, a process in which we must all take part.

Specific Proposals for an Ecologically Sensible City

Involve future users in the design process. This can be either informal, as in the South Bronx, or formal, as Alexander has demonstrated. He worked with a representative committee of users of a music school – secretaries, students, teachers – for an entire week, starting with stakes in a muddy field and ending with a detailed design. Co-housing is another, more formal, example. Land use committees would require user participation in any project's design before they would approve it.

Encourage small scale projects. Specifically,

- No building more than four storeys high
- No more than 2740 square metres total indoor area
- No more than 900 square metres to a storey
- If more than one building is needed, the buildings should be connected by arcades, paths, and bridges

Mix land uses. Specifically, buildings should be permitted to house different uses – offices, workshops, residences, shops, galleries, greenhouses, or recycling operations, all of which can co-exist if small enough and properly planned. If a building is devoted entirely or primarily to one use, proposals for adjacent new development must be primarily for different uses.

Concentrate land use. In terms of residential densities alone, this could mean anywhere from 75 to 200 persons per net hectare; however, since land use will be mixed, conventional density calculations are not much use. The main point is not to have any more single family dwellings at twelve to the hectare, or high rises that stack up hundreds of people to the hectare. The healthiest floor space index (square metres of floor divided by square metres of the lot) would be between one and two.

Encourage self-sufficiency. Developments would be required to be self-sufficient from the point of view of energy and waste disposal (both solid and liquid). The technology for this has already been referred to. Land use committees could also be instrumental in finding government-owned land as sites for small scale sewage-purifying systems and waste/recycling depots to service existing development. Decreased costs of garbage disposal would make money available to buy small sites and to staff them at the block level. Municipal utilities can work with the land use committees by allowing landowners to capitalize energy payments and use the lump sum to install earth-source heat pumps and solar collectors, to invest in small wind power installations, and to weatherize houses.

Discourage private cars. Cars should be actively discouraged by land use decisions. For example,

- Disallowance of parking as part of the development design
- Acceptance, and encouragement of, densities that would result in congestion if cars were the main means of transportation
- Favourable consideration of proposals to narrow streets and to widen sidewalks.

BIBLIOGRAPHY

Christopher Alexander. *The Timeless Way of Building* (New York: Oxford University Press, 1979).

——. et al. *The Oregon Experiment* (New York: Oxford University Press, 1975).

Wendell Berry. *The Unsettling of America: Culture and Agriculture* (San Francisco: Sierra Club Books, 1986).

Edmund P. Fowler. *Building Cities That Work* (Montreal: McGill-Queen's University Press, 1992).

Roberta Brandes Gratz. *The Living City* (New York: Simon and Schuster, 1989).

Dolores Hayden. *Redesigning the American Dream: The Future of Housing, Work, and Family Life* (New York: W.W. Norton, 1984).

George McRobie. *Small is Possible* (London: Abacus, 1982).

Ruben Nelson. *The Illusions of Urban Man* (Ottawa: Square One Management, 1979).

Oscar Newman. *Defensible Space* (New York: Macmillan, 1973).

Bernard Rudovsky. *Streets for People* (New York: Doubleday, 1969).

John Seymour and Herbert Girardet. *Blueprint for a Green Planet* (New York: Prentice-Hall, 1987).

Bruce Stokes. *Helping Ourselves: Local Solutions to Global Problems* (New York: W.W. Norton, 1981).

John Todd, with George Tukel. *Reinhabiting Cities and Towns: Designing for Sustainability* (San Francisco: Planet Drum, 1981).

Reflections on *Building Cities That Work*

The cities of North America are not just ruining the environment, they are ruining us. The fact should not have to be pointed out, since we are symbiotically linked to our built and natural environments. If our environment is in trouble, then so are we. Yet in our minds we have made ourselves separate for so long that we must make a conscious effort to reconnect. Our cities must be as much a part of this process as our rural settlements and agricultural practices.

In *Building Cities That Work* I showed how destructive postwar urban development has been to the economy, to the environment, and to our social fabric. Spread-out cities are expensive, due in part to our dependence on the automobile – a major reason why city air is so bad. Not only are municipal services such as water and sewerage more costly because of urban sprawl, but in general the large scale and segregated nature of our all-new urban development costs us billions in unnecessary dollars.

It is also now clear that our achievements – urban growth and, allegedly, a higher standard of living – have been obtained at the expense of astounding damage to the ecological systems that support life on this planet. Large numbers of mainstream scientists, not just crackpots, are telling us categorically that if the rate of environmental damage is not only halted but reversed in the next few years, we've had it.

What Has Happened to Street Life?

Since 1945, we have undertaken a massive urban and suburban construction program – out with the old neighbourhoods and in with the new. This growth period also brought unprecedented luxuries to many. However, it has not made us any happier, as numerous public opinion surveys show, nor has it improved the quality of our society, as evidenced by widespread violence and crime, overflowing jails, city streets that are shunned after dark, and numerous other indicators of an unhealthy social life. As some perceptive soul remarked, our standard of living has ruined our quality of life.

Urban sprawl, segregated land use, and large scale development have had frightening influences on our communities. There is now considerable evidence to show that postwar building patterns are associated with decreased contact among neighbours and increased crime, particularly juvenile crime. Children, in fact, are especially harmed by the built environment, which separates them from the adult world and denies them unprogrammed spaces in which to play or just to hang out. They never learn how to behave responsibly in public and semi-public places. Single-family suburban housing developments, not just inner-city neighbourhoods, are turning out juvenile delinquents.

The Death of Politics

Peoples' living patterns have become more private. Much of life is spent inside single-family homes (or in backyards); in high rise apartments far removed from the street; or enclosed in cars, getting to and from destinations around the far-flung city.

The politics of the postwar city show few signs of vitality. Politics *is* the mass media for 99.9% of the population. A minuscule proportion of us experience direct, authentic politics, which I would define as getting together with others in the community to discuss a problem and then doing something about it collectively. We look instead to elected representatives for solutions, who are glad to oblige with plastic promises delivered in ten second spots every few years, on the tube. Our political life, in other words, consists of the passive reception of manipulated electronic images. This is not political life; it is the death of politics.

What has disappeared in the postwar city is a casual public street life, where nothing in particular may be discussed but through which the social and political vitality of a local place is maintained. This street life needs small-scale, mixed and concentrated land uses, described over forty years ago by Jane Jacobs in *The Death and Life of Great American Cities*. The physical form of our cities discourages casual sidewalk contact with neighbours and, with it, authentic politics. We do have a public life, of sorts: at the malls, on public transportation with thousands of others, and at work. Many of us work in situations where we interact with dozens or even hundreds of people every day. Unfortunately, our relations with them tend to be narrow – serving a cup of coffee, quoting prices on services or merchandise, giving an injection, teaching business skills.

We are less connected to our neighbours and to the ebb and flow of activity on the street. There's been a loss of simple sidewalk contact: casual but essential greetings as we pass each other, the occasional chat about potholes or a troublesome dog, the odd remark about our children or our parents. These contacts, as Jacobs points out, may not be significant in themselves, but they provide a context for solutions when problems do arise, such as increases in vandalism or burglaries, proposals for a new development in the neighbourhood, or a misfortune suffered by a local resident.

There are political implications too in the fact that our cities have been built for us. We do not take part in their design or construction, a lack of participation peculiar to the twentieth century in Europe and North America. At other times and places, all but the most important public buildings or opulent mansions were designed and often built by their users.

Why Are Our Cities Like This?

Upon reflection, it is evident that although we can point our finger at the economic policies of governments or at the actions of big developers, governmental and development policies reflect sets of values and ideologies held by each of us. In other words, our unworkable cities demonstrate to us our own lack of balance.

This is not say that we should not examine the historical processes that have produced urban homogeneity and segregated land use. The market system has made urban land into a commodity even though, in many senses, it is not. Land is bought and sold for prices that reflect its uses, and its uses reflect the prices for which it is bought and sold. This process, at times concentrated and feverish, discourages diversity. As the price for land goes up, more and more uses are excluded. Also, more and more money is needed to be a developer, resulting in bigger and bigger companies who build larger scale developments.

In fact, the growth and movement of many different kinds of large corporations have been important factors determining the shape of our cities. Before firms moved to the suburbs, or new ones established themselves there, in both cases to escape the unions downtown, the division of labour and the growth in the size and specialization of firms had already introduced a separation of the workplace from the home within the nineteenth century city. This separation of land uses produced homogeneous areas of the city that were devoted to manufacturing, warehousing, commerce, financial and legal services, and housing. The destruction of mixed land use was started, therefore, by the growth of firms into large-scale corporations and by those corporations' production patterns. Then, at the beginning of the twentieth century, zoning began to be introduced, separating different land uses from each other even more and giving a strong impetus to the allure of the exclusively residential neighbourhood. It is no wonder we have physically homogeneous cities.

We have paid a price for this homogeneity. Already mentioned are the exorbitant costs of urban sprawl, stemming from having to drive everywhere because land uses are not only spread out but also separated, and from greater spending on infrastructure such as roads and sewers. In addition, neighbourhoods lose a degree of self-sufficiency when single use zoning forbids even corner grocery stores. Gone too is the architectural memory of a region. A Victorian building serves as a reminder to inhabitants that their city or neighbourhood did not just fall from the sky but that it has a history, one of change but also continuity. Too often new developments deprive their inhabitants of any connection to the past or to other regions. It has even been shown that architectural monotony is one of the factors contributing to lack of intelligence, and to juvenile delinquency.

Are There Any Cities That Work?

Is there, then, a city that works, or at least some that work better than others? While it cannot be denied that the title *Building Cities That Work* invites that question, in a way it is the wrong one, for two reasons. First, it implies that at present cities are discrete, definable entities. Second, it also implies that a working city is an end-point towards which we must strive. Once that end is reached, then what? Both of these implications must be addressed.

If a city is "working" then we assume it can be distinguished in some way from its surroundings and that it has identifiable internal dynamics. We also assume it has a size measurable in acres and people. While our present cities do have distinguishable activity patterns, these patterns are practically impossible to define spatially or demographically.

When one speaks of a city in North America in the 1990s, one is usually referring to a region of scattered urban development that surrounds a more densely built-up core. A name such as Los Angeles or Toronto identifies the core and its developed region, but it has become impossible to specify (except by arbitrary political boundaries) where the "city" ends and something else begins. In reality, there are no clear edges, either to older, more densely-built areas, or to the suburban housing and industrial parks. With the way we develop our urban areas, I would hesitate to call it city building. In fact, we have rendered the whole concept of a city extremely ambivalent.

Whatever we used to call a city is so wildly different from what we now have that we could reasonably expect in the near future to be living in settlements that we would not recognize as "cities" if we saw them now. The point is that because modern cities have no straightforward physical definition – we can't tell where they begin and where they end – it is difficult to judge whether or not they work.

But the question should not be avoided. When people ask me whether there is a city that works, my usual answer has been that I only know parts of cities that work, where, for example, neighbours are vitally engaged with each other and with the rest of the city in productive and creative ways.

This is, literally, a partial response. What about a whole city? Would it be composed of nothing more than neighbourhoods that "work", all stuck together? No, just as a working community relies on a set of responsible adults whose relationships work (mostly!), and is greater than the sum of its parts, so a working city will somehow reflect a higher level of spirit and culture than that of any constituent neighbourhood. The same principle works in nature, where inter-nesting organisms and organs work in amazing harmony to produce more glorious organisms, from molecule to organelle to cell to human, to community to . . . The glory of the molecule remains.

Here is where the second implication fits in. In a definitely nonrandom way, life transcends itself, has been transcending itself for eons. The city that works would be a living city and thus no different. It would express the magnificence of truly working communities at yet another level of magnificence. It would have its own life force. It would not be an end-point, but a growing, self-transcending organism that fits gracefully into its region, adapting and changing continually. Humans have in the past built cities that approached this ideal far better than contemporary ones. Patrick Geddes and Lewis Mumford elaborated on this subject decades before we started talking and writing about Green Cities. And although the ideal may sound impossibly abstract, it implies some fairly specific things about cities that work.

First, it implies healthy relationships. To the extent that cities or settlement patterns represent a relatively intensified web of human relationships, one point is clear: our relationships, both public and private, are not working. They lack many things – intelligence, trust, integrity, and most of all love. We also seem to have tremendous difficulties creating boundaries and making transitions between public and private relations, one of the jobs than can be effectively managed only by a healthy local community.

Can We Plan from Above?

Our present city form stretches these already unhealthy relationships over too many square miles, and segments them in time and space into patterns of work, family life, recreation, shopping, and so forth. The examples of small scale community experiments that do work, such as co-housing and community greenhouses, demonstrate what happens when people have taken up the challenge of learning how to live together. This process has to start with healthy relationships between partners, between neighbour and neighbour, and between humans and the land beneath their feet. With their actions, these experimenters express the most profound truth about a city that works: it cannot be planned from above. There is no blueprint or template to fit. Rather, there are new possibilities to grow into.

Many people, as sympathetic as they might have been to the analysis in *Building Cities That Work*, have not seen these examples as solutions to the problems it raises. Our "solutions," it seems, must come from laws and governments, from new "policy frameworks." If we think about it, passing a law about something means that things are not working, that someone must be forced to do something she or he would not otherwise do. It's an indication of failure.

Here, then, is a second characteristic of cities that work: they work from below. This means that they cannot be legislated into being. They would be participatory in the most radical sense. We would all be involved personally in growing much of our own food, fixing or building houses, and fixing our relationships, rather than relying on a specialist to do it for us. At the very least, such reliance would be on someone whom we know on several levels, as a neighbour and friend as well as a farmer or a dentist.

Working from below also means attachment to the soil on which the city is built. Most North American cities are built on prime agricultural land that is now ignored, denied, and paved, with disastrous consequences. We have so damaged the land with pavement and pollution that we cannot drink the water, nor do we even know how to grow at least some of our own food. Instead, it comes to us from a land that is being ruined elsewhere by an awesomely anonymous and destructive agricultural "industry." We are profoundly separate not only from the growing of our food, but from the soil beneath our feet.

Rediscovering the Land

An essential characteristic of a working city would be a working consciousness of the land it sits on and the land of its surrounding region. That connection between cities and their natural environments has always been there, but perverted and ignored by postwar urban development. Any sensible rebuilding process must acknowledge that cities and their regions are inextricably linked, that our urban dysfunctions are destroying the countryside, and that processes and practices taking place in rural areas (such as clear-cut logging and industrial agriculture) are threatening a healthy urban life as much as urban practices are threatening surrounding regions. We also need to celebrate and nurture the tenacious tendrils of wildness that keep greening the city's margins, in places we forgot to pave, even growing through cracks *in* the pavement. These tendrils are precious reminders that the city cannot escape its natural setting. In addition to participating in urban agriculture, our daily activities could be, unselfconsciously, a part of the maintenance and regeneration of the region's ecosystem. This is a joyful social and cultural exercise, not a solitary and wearisome economic task, something we feel we ought to do.

Overcoming the Dismal Religion

In fact, it is worth pondering for a moment why many of us are at present living our lives doing what we think we ought to do, putting off things we really want to do. If we truly love our "work" we are considered exceptional. We obey laws of the market and of profit seeking, forgetting that we ourselves have created those laws, along with our peculiar brand of economics, which we see as a "dismal science," based primarily on the scarcity of money and of goods and services.

What a dreary construction of reality! Real economics involves the way we feed and house and clothe ourselves, something countless cultures have managed to do in a healthy, integrated way all over this spectacularly abundant planet. Growing food and building houses are woven into their social and cultural life and are not considered particularly unwelcome tasks. As a result, food is more wholesome, houses and clothing are beautiful as well as sensible, and people show few of the chronic and social diseases of Western Europe and North America – until the intrusion of "economic development."

This is a curious religion that instructs its adherents to build polluting and dehumanizing factories, to destroy good soil with an agriculture of machines and chemicals, to undermine local self-reliance by trying to produce goods for fickle export markets, to dam rivers that wipe out towns and working ecosystems and replace them with stagnant, silt-filled water, and in general to ruin the land and local communities, all so that these adherents can make money. Perversely, once this form of economics is set in motion, no one ever has *enough* money, be it the developers, who are making truckloads of it, or the local workers, who are starving to death because now they need money to buy food they used to grow and because they haven't been paid sufficient wages.

No wonder we see economics as dismal. It is a separate, disliked activity that nevertheless rules the rest of our lives. Under its influence, we end up with no affection for places and even less for each other.

Cities, as concrete expressions of our dominant values, mirror both the narrowness and the meagreness of this religion. Vibrant neighbourhoods and businesses are routinely destroyed to make way for vapid retail complexes and immense expressway systems; productive soil is plowed under for oceanic parking lots surrounding yet another mall of identical stores or a WalMart. All this in the name of things we ought to do.

Local settlements maintained and cared for by ourselves would ensure an intelligent connection to the land, a connection which by definition is emotional as well as intellectual. The miracle is that intelligent, emotional attachment between humans and the land engenders similar attachments among humans. Our culture, our economics, and our politics weave together again.

This is the inescapable link referred to at the outset of this essay: our relationships to each other reflect our relationships to our environments – what we've built, the rural landscape, wilderness. Being close to each other and being close to our environment are practically the same thing, as mysterious as this may seem to some corporations who do not, who cannot, comprehend the sacredness of place to the human psyche. As institutions, corporations, which by their legal definition exist only for themselves, understand nothing but exploitation of humans and of places.

*

We have created massive built environments of concrete and steel that affect us in ways we are not willing to understand. It is also clear that changing things around is not only possible, it's something we must all engage in. What perhaps isn't quite so clear is that this project will be a pleasurable one that takes us away from our depressing preoccupation with a separated economics. It will be a joyful awakening to what we humans are really capable of. It's not just an opportunity to recreate economics, to redefine politics, but to grow our own culture, in every sense of the term.

Taking responsibility for our settlements – and for our food and our relationships – is a political process. Wendell Berry has remarked that one cannot be politically free if one has no control over one's food supply. Indeed, our vaunted political freedom is a cruel illusion as long as we depend on others to build our houses, to create our jobs, or to provide us with entertainment. There are some amazing people who have emerged from this economic coma and shown us that taking care of local places is the precondition for creating social justice and real democracy. (See Berry and Nozick.) Democracy and justice are empty shells if granted from above; they only take on meaning when they are filled with the stuff of our daily lives – the creation of buildings, the growing of food, the loving of one another.

BIBLIOGRAPHY

Wendell Berry. *The Unsettling of America: Culture and Agriculture* (San Francisco Sierra Club Books, 1977, 1986).
——. *Home Economics* (San Francisco: North Point Press, 1987).
——. *What Are People For?* (San Francisco: North Point Press, 1990).
Marcia Nozick. *No Place Like Home* (Ottawa: Canadian Council on Social Development, 1992).

The Politics and Ecology of a Healthy City

Our personal and public lives are connected in every way to the form of our cities. For example, there is convincing evidence that the layout of streets and buildings can support healthier public social contact and lower crime rates. Furthermore, it is increasingly clear that the pattern of urban development since World War II has produced, directly and indirectly, enough environmental degradation to threaten seriously our physical health – thousands die every year from car exhaust.

The evidence of environmental destruction is now written about so widely and reported on so frequently in the mainstream press that reactions are shifting from expressions of alarm to yawns. In the early 1990s we were told that the economic depression was a more pressing issue, for instance. With a few exceptions, nobody seems to be aware that our economic problems are environmental, and vice versa. It is all part of our curious and perverse culture of separation, habits of thinking that isolate problems into discrete categories and that cast us in the role of independent actors on the Earth.

There are, it is true, dozens of books about reconnecting to the Earth; but this message has become a catechism, whose relevance to our daily lives escapes us. In a real sense, to talk about reconnection to the so-called natural world is pointless, since we are already a part of it. Our cities are dysfunctional precisely because we ignore the inescapable symbioses between built and natural environments, and between ourselves and those environments. All the articles about reconnecting to the Earth notwithstanding, we are still advocating land use policies and leading personal lives that create ecologically corrosive cities, to say nothing of the damage we are doing to the countryside.

A healthy city form is possible, but reaching it is an adventure in wholism, in becoming aware of the many ways our lives and environments and problems are connected, not separated. This project can be both liberating and frustrating. It is frustrating because we are used to thinking in terms of one-way causation; this is how our science has worked to send men to the moon and to grow bigger ears of corn. While modern analysts have been working on dynamic models that allow for reciprocal causation, our Newtonian mindset gets confused when responses to stimuli appear to be simultaneous, a common enough phenomenon in ecosystems and in quantum mechanics.

Wholism is liberating because it doesn't matter where we start. Anything is fair game – at the individual or institutional level, in housing or agriculture. The argument of this essay is, first, that seemingly separate problems of urban form are connected, as are their solutions; and, second, that these solutions all involve becoming more aware of and participating in the physical environment of our cities. Actually, it is a mistake to restrict ourselves to urban policy problems. We think of cities as facing crises and dilemmas that are different from and unconnected to those of rural areas. We acknowledge the conventional interchange between the countryside and cities – raw materials and food, in exchange for mass culture and manufactured products – but we ignore the interrelationships between urban and rural dysfunctions.

Everyone will have a different label for those dysfunctions. In order to show how useful a holistic approach can be, I shall specify a few of them, just as an informal framework:

- Cities pollute the air, water, and soil for hundreds of miles around.
- Despite decades of unprecedented construction activity and production of goods, millions of people go without food and housing because they haven't enough money.
- Although statistics are notoriously inaccurate, crime is perceived as a serious problem in North American cities.
- Economic depressions and widespread unemployment continue to plague cities as well as rural areas.
- Most of us feel powerless to solve the above problems.

The relationships among these problems are such that misguided attempts to solve one of them can aggravate others, while an intelligent solution to the same problem may go a long way towards alleviating another seemingly unrelated problem. In the latter case, we produce more benign urban forms than the ones we presently have, so that their impact on our lives is healthier. Urban form, in other words, is part of the problem and part of the solution, cause and effect.

One of my favourite examples is given by Roberta Brandes Gratz in her book *The Living City*. She tells the story of Kelly Street in the south Bronx, which, in the mid-1970s, fitted the stereotype of that part of New York City – derelict buildings, vacant lots filled with rubble, and serious street crime. The street had a number of determined people, however, centred around Frank Potts and his family of eight. Over the previous fifteen years they had scraped and saved and worked incredibly hard to buy and renovate first one, then two, then more of the four story buildings on Kelly Street. When three more apartment buildings were scheduled by the city for demolition because of nonpayment of taxes, the Potts family, some neighbours, a social workers, and a few other helpers offered

> to take the buildings off the city's hands and renovate them. They were willing to . . . provide some unpaid labour. They wanted to build low-cost cooperative housing that would not be a permanent

> burden on taxpayers, as was massive subsidized new construction . . . They knew . . . that there was an endless resource of neighbourhood people looking for just this kind of job opportunity . . . Such a proposal was anathema to traditionalists . . . Why bother with such a small and complicated effort when many more, mass-produced projects could be built on vacant land, much more easily? (Gratz, 113 - 4)

Eventually, by seeking out contacts and support from conventional politicians, university and nonprofit groups set up to assist such ventures, Kelly Street was rebuilt by its residents.

The initial impetus was the threat of housing demolition; but, significantly, even while details were being worked out, the community started a garden on one of its vacant lots. Some time later, they organized a food cooperative and started recycling paper and glass for income. And, of course, dozens of young residents learned the skills needed in carpentry and construction for the renovation process. As new residents filled up the apartments, local merchants stopped going out of business and new stores opened. All this happened rather gradually, over a period of ten or more years.

Now think of the list of urban problems just outlined. The organization named Banana Kelly (after the street, which had a slight curve to it), produced low cost housing, grew some of its own food, stimulated economic development, and conserved energy with its renovations. Although I have no specific figures on the changing amount of crime in the area, certainly many youths were no longer unemployed, and unemployment is a central factor in explanations for inner city crime. Also, it might be noted that neighbourhoods with community gardens tend to have lower crime rates, other things being equal (Stokes). This happens not because of some "policy" to crack down on young offenders or to send out more police, but as a by-product of a neighbourhood that is socially and economically healthy – the two go together. Politically, the sense of powerlessness is gone, and a new self-confidence emerges, not in conventional political participation (such as voting), but in local action.

Finally, picture the kind of urban form that emerges from the Banana Kelly experience. It is relatively small scale – the buildings, the gardens, the street itself. The overall impression of architectural chaos gives way, on closer examination, to harmonious complexity. In spite of relatively high density, people are growing their own food. And it is not static and unmanipulable; Kelly Street will continue to change and grow, much like ecosystems undisturbed by humans. People in such a neighbourhood, as the evidence suggests, feel more connected to each other and more aware of their physical environment, because they are participating actively in shaping its character.

My second example is not a case study; it concerns one of the most basic elements of our lives – food. In *What Are People For?* Wendell Berry describes eating as an agricultural act. And agriculture is profoundly related to urban form, since where our food is grown, how it is grown, how it reaches our homes and in what form are all processes that involve buildings, roads, "open space," parking lots, watersheds, and, in general, diverse land use policies in the city as well as in the country.

The crisis we are facing here is not a derelict street, but a derelict environment. It is not an exaggeration to say that the way we feed ourselves (or, rather, feed some and starve others) is producing deserts both in the city and in the country.

While there are effective ways of meeting this crisis in rural areas, where new and old methods of organic agriculture show hope of reclaiming the soil, it is the city's land use and eating practices that we are concerned with here. Most cities sit on excellent agricultural land and are gobbling up more with wasteful and land-intensive suburban development. This stupid process can be countered with sensible tax policies that encourage higher density and infill development.

Just as important, when city dwellers start growing their own food, they contribute to the solution of a number of problems. First, they give less support to a system of socially and ecologically destructive agriculture. Second, they are less likely to go hungry, especially as they grow more skillful at city farming. Third, as mentioned above, community gardens are created by – and create – neighbourhoods whose informal public street life is healthier, resulting in less crime. Fourth, growing one's own food elicits a keen interest in where other food comes from and often results in initiatives to gain more knowledge and control over the food supply, such as community shared agriculture – food clubs buying directly from a local farmer. Or, gardening groups will start talking about housing, or day care, or drugs and will start working on those problems. It is easy to see how political and economic self-reliance can emerge from such a process.

While the landscape of a city that grows its own food may not look all that different from the air, the implications for city form are clear and far reaching: fruit bearing trees instead of ornamentals, orchards used as public parks, commercial farming on larger pieces of land, grapes and beans on vines in public squares, infill greenhouses in downtown backyards, and rooftop food production – which may indeed show up from the air.

How Do We Get There from Here?

One thing is sure: government policy cannot create a healthy city form. We put far too much energy into complex diagrams, guidelines, goals, and visions for the future. A good example of the policy mindset trying to grapple with the organic intelligence of the Kelly Street successes is provided by Gratz:

> Robert F. Wagner, Jr., like other public officials, was clearly in awe of efforts like Banana Kelly, yet puzzled about how to learn from it. "Yes, it is simply marvelous," Wagner said, "but the real question is, can we replicate it?" . . . He was asking the wrong question. . . . The revital-

ization of cities is not a science, but an art. It depends on creativity, variation, innovation. . . . For . . . "experts," solutions must be reduced to an exact repeatable formula. [C]ities cannot be approached this way . . . An urban organism – a block, two blocks, a neighbourhood, a whole city – must regenerate naturally to endure. Things must be allowed to happen and in a manner appropriate to a particular place . . . Each neighbourhood will be different, but each will have its own strengths, its own particular characteristics and people, and that will be its basis for future growth . . . As this happens, each improving neighbourhood becomes a visible example of what other neighourhoods can do, stimulating further renewal. This way, each neighbourhood replicates itself, instead of others replicating it. If government chose to significantly aid and nurture this process, a more significant renewal would have a chance. (41-2)

But such "significant aid" should not be thought of in terms of money. We still really think that money will solve most of our problems. In *The Death and Life of Great American Cities*, Jane Jacobs describes how what she calls cataclysmic money ruins rather than helps neighbourhoods. Similarly, Paul Hawken, author of *Growing a Business*, notes that many new businesses fail, not because of a lack of capital, but because of too much – people think that money can take the place of hard work. In fact, a community organization near Banana Kelly fell apart because its wonderful but well-publicized success attracted so much government money. Obviously, money is needed, but by treating it as a primary cause, we go astray.

We also must avoid plastic language and labels that convince us that we've solved our problems already. Pronouncements such as "We have a 'local development initiative' that will 'create jobs' and give the economy a 'kick start'" may be the first sign that we are losing touch with what makes community efforts work. Development has to be done by particular people in particular places. There is no single blueprint that applies everywhere.

There Is Plenty to Do

Bearing in mind these caveats about government policy, too much money, and plastic language, change can be imagined on three levels – attitudes, institutions, and yes, even policies.

The examples given above make it clear that nothing is going to happen without a widespread change in individual attitudes and behaviour, to say nothing of awareness. Such changes are indeed evident, in concern over the environment and in attitudes toward work, personal transformation and community involvement. It must be stressed, however, that it is difficult to discern just how much our personal lives are woven into a vast web of support for transnational corporations and big governments. Indeed, the literature on globalization is daunting. It suggests that it matters little what we do as individuals, since massive and increasing flows of capital, goods, and financial services and instruments are controlled by a very few corporations and individuals.

The more aware we become of our connections to natural and built environments and to each other, the more our behaviour will thwart corporate dominance over our lives, corporations being institutions that are essentially non-place-oriented. Thus, a second level of change is institutional. Institutions grow out of habitual patterns of behaviour, and the pattern at issue here is one of taking responsibility for urban land and form, in particular the urban and suburban land right outside many of our front doors. Christopher Alexander and his associates suggest the most sensible institutional change, from my point of view, in common land.

> Without common land no societal system can survive . . . In pre-industrial societies, common land between houses and between workshops existed automatically – so it was never necessary to make a point of it. The paths and streets which gave access to buildings were safe, societal spaces and therefore functioned automatically as

> common land ... But, in a society with cars and trucks, the common land which can play an effective social role in knitting people together no longer happens automatically. Those streets which carry cars and trucks at more than crawling speeds definitely do not function as common land; and many buildings find themselves entirely isolated from the social fabric because they are not joined to one another by land they hold in common. In such a situation common land must be provided separately, and with deliberation, as a social necessity, as vital to the streets ... The common land has two specific social functions. First, the land makes it possible for people to feel comfortable outside their buildings and their private territory, and therefore allows them to feel connected to the larger social system – though not necessarily to any specific neighbour. And second, common land acts as a meeting place for people. (337)

Common land is a place-oriented institution and is really the basis for the most elemental form of local government. Alexander et al. propose that, after working out local definitions of neighbourhoods and communities, users (residents and workers) take collective responsibility for the common land on their block or blocks; and community councils for neighbourhood clusters of five to ten thousand have control over common land forming borders between neighbourhoods. Thus, not only does public space become a central concept in local institutions, but also those institutions have direct responsibility for that public space. The examples of Banana Kelly and urban farming suggest that a healthy city form (or countryside) emerges easily when the most basic economic activities – growing food, building shelter – are done collectively. However, without some common land on the block, this is an uphill struggle.

The collective decisions of such institutions would obviously bear little resemblance to what we now call government policy. Still, one could imagine some of these decisions and suggest that central governmental institutions consider their merits. While resistance to such suggestions would be massive, they do serve to show in how many different ways we could move ourselves towards healthier and more conscious relations with our urban environment. The reader is reminded that we are dealing with a city form that reflects a whole culture of separation, which in turn is reinforced by the city form.

There are many reasons for the vacant lots that blight North American cities. Perhaps they reflect the fact that our own minds are vacant in places because we can't think of anything sensible to put there. In the case of the rubble-strewn lots, landowners are usually waiting for the market to make it profitable for them to build something. But the tax system helps them out. It allows businesses to subtract depreciation of buildings from their income tax. When the property is sold, the seller tears down the building to avoid paying capital gains tax. Hundreds of perfectly good buildings are destroyed because of this tax policy and hundreds more expensive new ones are built. In general, taxing buildings instead of land encourages speculation and raises the price of housing and retail space.

Following Henry George's argument that the community gives value to land and therefore that land, not buildings, should be taxed, Pittsburgh taxes land more than buildings. As a result, fewer good buildings are demolished and housing prices are lower than in other North American cities. This makes sense both economically and ecologically: renovating old buildings uses fewer scarce resources and more local labour than building new ones. Older buildings are also more likely to fit into the fabric of the city, although new ones could as well, if planners and builders were more thoughtful.

A second and major policy change should focus on the car. Since car use is a central factor in our cities' unworkability, which stems in part from wasteful land use patterns, adoption of various creative proposals to squeeze out cars would be a giant step towards the rational use of urban land. Our tax system subsidizes the car driver far more than the cyclist or transit rider. There are many remedies for this situation, most of them politically suicidal right now, since as noted earlier the necessary changes in attitude have not occurred. Here are some suggestions: a seven year moratorium on road and freeway construction; traffic calming schemes that slow traffic to thirty kilometres per hour or less (already being done in many cities); and license fees in the three and four figure range to cover the many secondary costs of the car, such as pollution.

Finally, all new development should be designed by users. There is clear evidence that houses and work spaces are more functional when residents and workers are completely involved in the layout and even construction of those spaces. This policy would produce smaller, environment-friendly buildings, and neighbourhoods that are cared for, with the accompanying advantages of a healthy street life, low crime, and economic self-reliance.

Policies such as these presuppose significant shifts in cultural values and priorities. We need attitude change more than new policies. Besides, government policies are inherently incapable of being holistic, which means that the nature of our problems and of their solutions cannot be fully addressed.

Housing policies, for instance, tend to focus on the writing of new laws about or the appropriating of money for the construction of housing in general. This separates the problem (if there is one) from its place, which connects building houses with other activities. By separating problems from places and from each other – housing from unemployment, for example – government policies make those problems more intractable. In any case, as Gratz points out, what works for one time and place won't work for another.

In their study of local power politics, *The Sustaining Hand*, Bryan Jones and Lynn Bachelor show how we create a solution set – a constellation of ideas, procedures, and networks of contacts (which often turn into bureaucratic structures) – to solve a particular problem and then blindly use the same solution set in other places and at other times, no matter how inappropriate it is. And we scratch our head because things didn't work out.

The policy mindset, which separates places and actions, goes to the heart of why we are destroying our ecosystem. It is the same blindness that feeds our illusions of independence from the Earth, from our own built urban form, and from each other. But, as some of my examples show, many of us are rediscovering the wondrous symbiosis which is really there, and which is no illusion.

BIBLIOGRAPHY

Christopher Alexander et al. *A Pattern Language* (New York: Oxford University Press, 1977).

Wendell Berry. *What Are People For?* (San Francisco: North Point Press, 1990).

Roberta Brandes Gratz. *The Living City* (New York: Simon and Schuster, 1989).

Bruce Stokes. *Helping Ourselves* (New York: W.W. Norton, 1980).

It Accidental or Merely Wild?

There are no accidents. Don't confuse me with the facts; nothing can sway me from this belief.

Things happen when we don't mean them to happen, and we often call such occurrences accidents. Or, we may observe a coming together of people and events, which on the surface seems highly unlikely or quirky, and call that an accident. But it isn't. The trick is to explain things, remembering that what works as an explanation for some doesn't work for others.

Robert Fulford has written a book about Toronto called *The Accidental City*, which purports to describe notable features of that city as the outcome of a series of accidents. Let's take Fulford's example of Hurricane Hazel, whose floods killed over 80 people in Toronto in October 1954. This "accident of nature," writes Fulford, brought Torontonians to their senses; and they started reclaiming Toronto's ravines from human development so that their rivers and streams could flow freely if they needed to, without danger to human life.

There is no doubt that Hazel raised their consciousness, or that Torontonians subsequently developed a healthy respect for the ravines' ecological niche in their landscape. Neither is there doubt that this respect blossomed into a celebration of the ravines and watersheds as special, wild places that form part of Toronto's identity.

Hazel was no accident, however; nor was her passage through Toronto. Sooner or later, a rainstorm of her size and intensity would have delivered nature's message: do not build, or pave, too close to the ravines, or human lives will be lost. In fact, civic leaders did the only sensible thing, reasoning that the deaths were the result not just of heavy rains, but also of where and how Toronto had built: they returned ravines and floodplains to their natural state (mostly). Treating something like Hazel's destruction as an accident would have been an excuse not to take responsibility.

Being responsive to these messages, which remind us that we are part of energy patterns far beyond our own awareness, connects my response to Fulford's book with another recent book on Toronto, by Wayne Grady: *Toronto the Wild*. Although Fulford's book seems on the surface to concentrate on the built environment, and Grady's on the stuff in between, both address the interface between the human-built city and its ecosystem.

Fulford, a respected art critic, journalist, and long-time editor of *Saturday Night*, provides a stimulating string of unconnected chapters on what used to be called Metropolitan Toronto – now the single megacity Toronto, after being amalgamated by Ontario's Tory government in an act of breathtaking simple-mindedness. Although Fulford's theme is given in the title, and although he describes a number of (so-called!) accidents, the work has no other ostensible thesis. It is, at the same time, an informed and intelligent exploration of selected features of Toronto's history, mostly recent.

This is as it should be, one might argue. The city is a complex phenomenon that can be observed from many angles and in many dimensions; the book simply reflects the truth, adding a teasing, footnote-like sentence on the last page, to the effect that the city is never a finished product – it is always being built. However, I feel that all this fascinating material could use some sort of framework that pulls it all together, and an author as talented as Fulford could be expected to deliver more.

The framework of "accident" is not convincing. The Oxford Dictionary defines an accident as "an unforeseen occurrence" or one "without apparent cause." Can today's Toronto really be so characterized? Most of Fulford's accidents seem to be of the unforeseen variety: the Henry Moore sculpture collection at the Art Gallery of Ontario; the emergence of the CN Tower as a tourist attraction and as a central image for the city; the preservation of Toronto's ravines as a series of interconnecting parks; the arrival of Jane Jacobs in 1969; the forlorn and unimpressive North York Performing Arts Centre (subsequently called the Ford Centre); and Toronto's motley array of public art, uninformed by any collective civic intelligence.

Consider the Henry Moore sculptures. Fulford carefully describes how the new city hall's Finnish architect, Viljo Revell, who admired Moore's work, visited the artist in 1964 and talked with him about creating a sculpture for the square in front of the city hall, actually selecting a working model. Revell died of a heart attack the next day, but the project was pursued from both sides of the Atlantic. Although the work caused much controversy in Toronto, Moore's connections with city officials and art collectors continued. "In Moore's conversations with various Toronto friends," says Fulford, "the idea of a Moore donation to Toronto came up." The subsequent negotiations and planning are traced, indicating that, although the final outcome was never foreseen when Revell visited Moore, an intricate dance of personalities and events combined to produce the Henry Moore Sculpture Centre, "the most important public collection of Henry Moore's work anywhere, and the largest single gift in the history of Canadian museums." The pattern of the dance may not have been clear to the dancers, but here was a pattern nevertheless.

Fulford describes similar processes with the CN Tower, which was supposed to have been the centrepiece for a massive, multi-billion dollar development in downtown Toronto on lands owned by the railways, not a stand-alone symbol of the city. Like Moore's sculptures or the North York Centre for the Arts, the fate of the CN Tower is more accurately explained as a consequence of deliberate, identifiable decisions. Those individual decisions can be explained quite rationally, just as Jacobs' decision to move to Toronto can: "She and her husband left New York because they found America in the Vietnam era repressive – and because they had two draft-age sons," writes Fulford. In addition, while they could have picked any city, Toronto seemed an "agreeable place." And, besides, it offered work for Bob Jacobs, who was an architect.

These events did not happen by chance. The author describes the causes and contributing factors. They were unexpected, however, and that seems to qualify them for the other half of the Oxford definition. But consider the nature of the unexpected: it simply means that the interested observer, or the involved participant – or even the uninvolved and uninterested person – is unaware of or insensitive to the web of events and contributing factors producing the outcome. Things are therefore seen as unexpected, and, therefore, termed an accident. The unexpected can be defined, then, simply as something with no apparent cause, with the emphasis on apparent. With the 20/20 hindsight of a gifted storyteller like Fulford, the unforeseen becomes comprehensible.

Bear with me. I am going through all this because much of Fulford's book deals directly with urban planning, an intentional exercise which has produced many unintended consequences, and it behooves us to ask why. These unintended consequences were not accidents, because as Fulford himself shows, they did not occur by chance. Calling them accidents is just an admission that we don't know why they happened. I am pleading for us to be more aware of how we fit into the cosmos, as we pursue our own conscious intents.

Indeed, *Accidental City* is full of examples of planning and city building carried out with varying degrees of awareness. The chapter on Fred Gardiner describes how this man, the first chairman of the Municipality of Metropolitan Toronto, created the expressway that bore his name, along with billions of dollars' worth of other concrete infrastructure that encouraged the use of cars and the building of Toronto's suburbs. Gardiner's city, it could be argued, was definitely not accidental: Fulford summarizes the usual argument that Gardiner was a skilled politician who was not afraid to use threats and pull at Queen's Park to further his building program. He was given awards citing him for his outstanding leadership.

Times have changed, and for some 20 years there have been noises made about tearing down the Gardiner Expressway so that Toronto can reconnect to the lake. But, writes Fulford, that would take the kind of leadership illustrated by Gardiner himself, and that is not a likely prospect.

Fulford's own account, however, shows that Gardiner's construction frenzy was widely supported by the public and the press of the day. A similar point has been made about Robert Moses, who has been widely denounced for his brutal destruction of neighbourhoods in New York to build expressways and high rise housing developments. But in fact he was concretizing, literally, contemporary conventional wisdom about city building. In Fulford's words, Fred Gardiner and his allies were playing their part in one of the great dramas of urban history, the transformation of the North American city from a single unit into a loose confederation of regions linked by superhighways. The old-style city placed the various activities of the people close together, but the new post-1945 confederation separated them. This was the moment when city builders embraced the modern idea of specialization and tried to make urban planning as rational as manufacturing.

That Gardiner was a flamboyant reflection of the collective

will of the era doesn't make his policies less stupid; the point is that they were no accident – just insensitive.

Fulford's negative evaluation of the specialized city, like my own, is derived from Jane Jacobs' perceptive observations about how social as well as economic vitality thrives on small-scale, close-knit physical diversity in cities. Economic vitality should not be equated, here or elsewhere, with the accumulation of dollars, any more than social vitality should be equated with more people. Jacobs shows only too clearly how what she calls cataclysmic money is disastrous to the healthy fabric of the city.

Scattered new developments pay lip service to her theories, and we are now vaguely aware that our life support systems are put at risk by urban sprawl. But the sad truth is that despite widespread and articulate critiques of suburban development and urban expressways, and despite continuing in-your-face evidence that this development is awesomely expensive, it is still the norm. Governments claiming to be petrified of deficits continue to finance infrastructures and building patterns that cost those same governments billions of unnecessary dollars in everything from health and social services to economic recessions and environmental destruction. Anyone doubtful of these facts should consult Jacobs' books, Pamela Blais' background study for Anne Golden's Report on the Greater Toronto Area, or my own book, *Building Cities That Work*. Everyone in North America is studiously ignoring the extravagant costs of postwar development. This idiocy will continue until an "accident" like Hazel will make it even clearer that we need to change our living patterns, not just public policies.

Lack of awareness lies behind other accidents described by Fulford, but it needs to be asked why some of them are considered happy coincidences and others just blunders. One of the opening themes of *Accidental City* revolves around the importance of public squares, and in particular Nathan Phillips Square in front of Toronto's City Hall. Fulford quotes Spiro Kostof, who points to

> the universal human need for public space: "Cities of every age have seen fit to make provision for open places that would promote social encounters and serve the conduct of public affairs."

The square (continues Fulford) is a real triumph. Since opening day, it has been the great living room of Toronto, the place where citizens gather to hold public meetings, to celebrate triumphs, to mourn lost heroes like John Lennon or to welcome great figures like Nelson Mandela.

Fulford sees the creation of Nathan Phillips Square as a coming of age for Toronto's civic consciousness. And, to give him credit, he returns to this theme of Toronto's going public when he writes about "The Beach" (called "The Beaches" by the rest of Torontonians), a neighbourhood transformed in the 1970s from an insular, Anglo-Saxon inner suburb, into a trendy area with many cafes and a lively sidewalk culture along the main drag, Queen Street East. The Beach perfectly illustrated how Toronto was developing a public street life, according to Fulford.

But all this happened in an older section of the city, like Nathan Phillips Square. The theme returns a few chapters later, when he begins his treatment of "downtown" North York by stating that "all towns of any size eventually realize that public spaces are vital to their psychological well-being, which in turn is crucial to their economic health." However, Fulford pans Mel Lastman Square, named after North York's all-but-eternal mayor, calling it "cluttered and incoherent."

Now the relevant question here is, What lessons can we learn from these three examples of going public? Jacobs considers the reasons for the success of parks and squares in her book *The Death and Life of Great American Cities*. She argues quite sensibly that parks and squares can never become by themselves successful foci of neighbourhood civic life unless they are surrounded by lively, physically diverse streets. That is, each part of the city is symbiotically related to its surroundings. Jacobs' condition is reasonably fulfilled in the case of Nathan Phillips Square, and not at all in the case of Lastman Square. Queen Street East is not a square, though it is bordered by a park; but the street's vitality is supported by short blocks, a mixture of old and new buildings, concentration of uses, and a mix of land uses – Jacobs' definition of physical diversity.

Fulford's diagnosis of Lastman Square does not refer to his other examples of public life in Toronto. Instead, he writes, "You could stand in what purports to be the core of it all, Mel Lastman Square, for a long time before the word 'planning' entered your head." This is a strange sentence for an admirer of Jane Jacobs, an ardent opponent of most formal city planning.

The paradox is made all the more clear in Fulford's account of the St. Lawrence Neighbourhood, an area developed by Toronto reformers in the 1970s. This redevelopment may have been an accident in Fulford's eyes because Jacobs' presence here was (to him) a stroke of luck, and she had much to do with the new design. He calls it a miracle, since it was a hugely successful example of a planned-all-at-once neighbourhood with social housing, yet built at a time when such developments were seen universally as a mistake. Some accident! As the story unfolds, Fulford shows how the determination of a few intelligent politicians, planners, and citizens (including future residents) brought the St. Lawrence Neighbourhood into existence. Fulford's story even includes an accurate summary of the connection Jacobs makes between physical diversity and "the spontaneous inventiveness of individuals." It is worth noting that the St. Lawrence Neighbourhood was revived with the active assistance of Jacobs, the most famous and articulate critic of preplanned social housing projects.

But planning for spontaneity is a contradiction in terms, whether it's a square or a neighbourhood. In Jacobs' view, if you plan for it, it goes away. I call this principle the Zen of Planning. However, this does not mean that lively public street life happens by accident. It is a complex but clear reflection of the symbiosis between physically diverse neighbourhoods and thoughtful, creative choices by individuals and small businesses who feel part of their neighbourhood. In other words, we can't plan for urban vitality, but we can allow it to grow.

The history of the St. Lawrence Neighbourhood is a history of trust in the creative possibilities of participation, a process which may seem messy to those of us who like to have things under control, but which has its own logic and patterns of energy, something Fulford seems to understand. That is why it's puzzling when he calls the outcome accidental, when in fact it was simply unpredictable. The really successful parts of Toronto illustrate the fact that when engaged, aware citizens at the street level are involved in collective action, the results cannot be specifically foreseen, but they always make sense.

I totally agree with Fulford that Jacobs has it right and that Gardiner's segregated city is a colossal mistake. His book is frustrating to read, however, because Fulford has not used his reasoning on such issues to weave the separate chapters into a powerful argument about what makes cities work, a subject he obviously loves. I maintain that much of what he recounts about Toronto's transformation exhibits a shocking lack of awareness on the part of city builders, not only about how parts of the city fit together, but about how the city fits into the biosphere.

This is where Wayne Grady has much to teach us. Grady's book is written in the tradition of Anne Whiston Spirn's *The Granite Garden* and seeks to describe the plants, animals and insects that inhabit the city of Toronto. Grady, a former editor of *Harrowsmith*, and author of books on dinosaurs and coyotes, reveals that many of our actions have indeed produced unforeseen consequences with respect to other species. He points out, for instance, that although we generally think of cities as eradicators of animal habitats, those same cities attract numerous other animals because of the abundance of food (pigeons, coyotes) or the creation of a convenient new habitat (ring-billed gulls). In the latter case, the Leslie Street Spit, one of Fulford's happy "accidents," figures prominently.

The Spit was created out of landfill, and even broken pieces of old buildings, from excavations during Toronto's building booms of the 1970s and 1980s. It is a curved peninsula reaching some five kilometres into Lake Ontario, enclosing part of the Toronto Islands. It started as an extension of the Toronto Harbour, but when new shipping failed to materialize, it was decided to let it grow as a recreational area. No one suspected how popular it would become for wildlife. "By 1980," writes Grady, "it had been colonized by 278 species of plants, from wild grasses and hedges to full-grown trees and shrubs." And birds. More than 250 species of birds have been sighted there, birds like "Caspian terns, black-crowned night herons, blue-winged teals, Hudsonian godwits," and "probably the largest ring-billed gull colony in the world."

The ring-bill had been slaughtered, like the passenger pigeon and many other birds, in the late nineteenth century. In 1973, only twenty-one nests were found in Toronto Harbour's eastern headland, the bump that was to become the Spit. By 1984, because the Spit recreated the perfect conditions for the ring-bill's nesting needs, there were 74,500 nests. I guess the word spread fast.

Toronto obviously did not set out to attract ring-bills. It is sobering to reflect that humans have been active, if somewhat inept, participants in shaping what flora and fauna surround them in Toronto, as Grady illustrates beautifully. Most of our plants are non-native, brought here both inadvertently and on purpose. We brought in the cute little English sparrow; then when it overran the place, we killed the pesky things by the thousands. Our urban construction has produced many square kilometres downtown of relatively warm, moist, sandy soil – a perfect environment for a certain form of termite. Although termites are tropical, and usually form new colonies by fleeing to new habitats, the termites in Toronto adapted themselves nicely to a life entirely underground. There is a serious termite problem: 18% of the houses are infested. As a northern city, Toronto is somewhat unique in this regard, although it is a uniqueness it sometimes feels it could do without.

Other chains of events described by Grady should be read as cautionary tales to remind us of the intricacy of the web of life we find ourselves in. One example concerns a type of mosquito previously only known in Asia, but identified in Ohio in 1988. It was traced to Houston, where it was found in used tires imported from Japan from retread companies. In 1989 (not 1988), this mosquito started to carry an encephalitis virus. Another example: the English sparrow population declined in North American cities because of the spread of the car in the 1920s: apparently the little bird's winter survival had depended on seeds from horse manure!

Grady's book is entertaining not only because he makes such interesting connections, but because he does it with a wry sense of humour. On the other hand, he gets terribly detailed, not only about different animals' behaviour, but also about the diseases and pathogens carried by raccoons, pigeons, cockroaches, and mosquitoes. I now know more than I ever wanted to about rabies and theories on the spread of encephalitis. But I acknowledge that other types of readers may really lap this up. In any case, the interrelationships between humans (and our buildings) and the plants and animals of Grady's book need to be explored more systematically by specialists. Grady covers himself, in a way, by subtitling his book as "Field Notes of Urban Naturalist," which excuses the fact that – like Fulford – he doesn't try to make regular connections between chapters. *Toronto the Wild* is an almost too conscientious treatment of a dozen or more animals and plants that surround the city dweller, but the treatments are isolated from each other, and only fitfully does the analysis relate to the meaning of this in the context of the city.

The sense one gets by the end of both books, however, is that if there is chaos in Toronto, it is from the behaviour of humans, not of termites and raccoons. Jacobs once remarked that attempts to create order with geometrically, large scale buildings actually boomerangs. Their monomaniacal regularity and size are telling the user "it doesn't matter where you go." The user becomes disoriented, because the builders have really produced disorder out of abstract homogeneity. It is, says Jacobs, a form of chaos. Grady shows how impressively plants and animals interweave their patterns of life into the city's margins, which we call waste places. Here there is order, but an order characterized by subtle and fluid relationships, by small and creative connections among the lives of weeds and gulls and termites – and humans.

Toronto the Wild is, if nothing else, an admirable catalogue of other organisms' sensible and even ingenious adaptations and inventions to environmental dislocations, often caused by humans. In this respect, Toronto the wild has something to teach Toronto the accidental: pay attention to Mother Nature. City building could take on a whole new meaning.

BIBLIOGRAPHY

Robert Fulford. *The Accidental City* (Toronto: Macfarlane Walter & Ross, 1995).

Wayne Grady. *Toronto the Wild* (Toronto: Macfarlane Walter & Ross, 1995).

Getting Urban Growth Wrong

Every once in a while someone comes along to tell us that we are at the end of ideology, or of history, or even of scientific discovery. In fact, scientific discovery itself has revealed to us an astounding story of unending development from an unbelievably hot concentration of particles to galaxies, stars, planets, molten rock, micro-organisms, plants, animals, even humans! To presume that this majestic evolutionary process has ground to a halt with us is arrogant.

Humans are, of course, as much a part of nature and its evolutionary flow as the oak tree and the amoeba; our functions and activities are completely at one with the rest of the river of life, especially at the cellular and molecular levels. Inside each of us are millions of organisms essential to our metabolic processes, to our survival, yet not attached to us like an ear or a foot; they are constantly passing through us and being replaced by millions more. By means of this process, most of the human body's cells are replaced yearly, or even monthly or daily. They break down into atoms and molecules, and circulate to other organisms. You have atoms in your body that have been in my own – and in the bodies of Jesus Christ, Napoleon and Emma Goldman.

Throughout this constant interchange, our outward physical appearance changes little, yet most of us believe we are able to "develop" our talents and capabilities, even radically change our outlook on life and living conditions. It seems that in the past we changed form as well: most of us believe that humans "developed" from primates, which in turn developed from other animals, plants, and so on.

These brief glimpses are offered as a starting point for reflection on the title of James Lemon's book, *Liberal Dreams and Nature's Limits*. Lemon, a professor of geography at the University of Toronto and an active citizen of one of Toronto's downtown neighbourhoods, tells us that dreams of never-ending economic and physical growth for cities, for our urbanized societies, have run up against the limits of nature. But his definition of these limits is only partly concerned with ecological degradation, which he summarizes mostly in non-urban terms such as fish stocks, soil depletion and exhaustion of mineral reserves:

> Less apparent – but at the exact centre of my underlying argument – is nature's subtle ability (as it were) to slow the growth of population and even the economy in a finite world . . . Growth can never continue at the same pace and eventually slows down. People are time-bound natural creatures. All attempts to annihilate space and time, to simplify tasks, like Sisyphus, reach limits. The harder we try, the greater the cost.

But nature itself is an energy that does not stop progressing. Some of its systems' principles may put limits on what we humans (who are part of nature) undertake, but nature provides us with one example after another of life transcending itself, from single-celled to multi-celled organisms, from termite colonies and their amazing collective feats to human societies. Lemon gives us a handsome example of such development by describing, often in great detail, the physical contours, the social life, the politics and the economics of Philadelphia in 1760, New York in 1860, Chicago in 1910, Los Angeles in 1950 and Toronto in 1975. For all their noxious impact on the biosphere, cities have indeed transformed themselves several times over in 250 years.

Lemon's definition of nature's limits, which is in terms of some ceiling on human technology, underrates both technology and nature. Our technology has transcended itself many times by creative thinking: as cars show themselves to be progressively less suitable for sustainable urban living we improve accessibility to work and shopping by intensifying land use. As far as nature is concerned, I feel that "constraints" need to be reconceptualized as what we humans can do without exterminating ourselves, not what nature can or cannot sustain.

The paradox is that we are part of nature, and yet we have been rendering our local environments uninhabitable for millennia. Why, as one of nature's creatures, with all this potential for self-transcendence, have we developed such a strange passion for self-destruction? The curious history of this passion goes so far back that one can hardly blame the liberals for our ecological stupidity, which is Lemon's tendency. However, liberalism's contribution to the problem is intriguing.

Lemon's characterization of liberalism, "in its guise of American conservatism," is grounded in a historical analysis of seventeenth century England and points to the centrality of individualism, especially as it related to property (land) ownership. As Lemon notes, the fever of most immigrants to North America to have control over their own plot of land came from being kicked off the land by the enclosures in Europe. This process, through which nobles and gentry took physical control of land in order to farm and raise livestock for profit, accomplished three things simultaneously. First, it turned land into a commodity; second, it created a large group of dispossessed people desperate to own land so they could regain some security in their lives; and, third, it made land into something to be exploited for private gain as opposed to a place whose natural features became part of its occupants and their culture. Along with the contemporaneous scientific revolution, which sought to study nature "objectively" in order to predict and control it, the enclosures produced generations of Europeans and North

Americans who saw themselves as somehow separate from nature. This could explain our disregard for sustainable development.

Lemon points out how the shibboleth of private property was used to justify not only the merits of the pioneer homestead and of home ownership in post-war suburbia, but also tolerance of unscrupulous speculation and the spread of huge development corporations. The liberal dream in his title, however, applies to all of the above, to homesteaders and homeowners as much as to corporations. Everyone, according to this dream (or credo), should be allowed to pursue material wealth at will, since the well-being of society could only benefit from continual economic growth fuelled by individual entrepreneurship. Lemon argues that North American (especially American) liberal ideology is centred on never-ending physical and financial expansion, and on newness, whether apparent or real. He also underlines liberals' distrust of government, which interferes in the inalienable right to do what one wants with one's property and to enjoy the fruits of one's labour. While there is much more to liberalism, of course, these elements are, to Lemon and to many of us, at the core of North Americans' liberal dream.

He makes a point of criticizing Jane Jacobs' prescription for healthy urban economies not because he disagrees with her vision, but because he thinks it is unrealistic, swimming upstream against this liberal ideology. Jacobs' thesis, in her classic book, *The Death and Life of Great American Cities*, was that economic vitality (and social well- being, in fact) comes out of concentrated and physically diverse city neighbourhoods with small-scale businesses. For Jacobs, big monopolies and governments, with their endless mega-projects and "cataclysmic" amounts of investment, spell economic stagnation, not growth.

My contention is that liberals, conservatives and socialists alike are not rigorous enough about defining the nature of growth and development. Much of what is called development is simple expansion: the same thing, only more of it. The evolution of life cannot be so characterized. Rather, evolution occurred because organisms regulated their own populations and interacted with the environment of other organisms to create increasingly complex interdependencies, to the point of creating whole new levels of life forms. Humans have taken part in this process of diversification in the past, but we seem to have reached the point of simple expansion, of doing more of the same thing only faster (the computer is a good example of this). We still call this progress or growth. The result is that we are hitting a wall, not so much of nature's limits as of our own mindlessness.

This distinction between types of growth explains much of the difference between Lemon and Jacobs. Lemon sees nature as putting limits on the same old kinds of growth; Jacobs sees diverse city neighbourhoods as the natural source of new kinds of growth (economic, social, political).

For the most part, then, Lemon labours under the old definitions of growth and progress – jobs, production, income, income equity, urban infrastructure, clean affordable housing, sanitation, social services, and so on. Now and then he expresses some telling and intelligent misgivings about conventional concepts of progress, calling it at one point nothing but "offices and the Olympics," and at the end of his book proposes "creative construction and conservation." He is also decidedly skeptical about the real novelty of computers and of the so-called information revolution.

However, rather than stimulating him to rethink the meaning of urban growth, these misgivings reflect a pessimism in Lemon, who worries about "how to maintain civility and equity in an economically stagnant world," and argues that cities have become parasitic. It is almost as if he has given up, not just on human progress but on evolution as well. Interestingly, his book has many examples of how cities have defined new levels of civility and economic growth in the past, examples of connections between the search for public good and the search for private gain. Cholera epidemics and fires that destroyed whole districts of New York in the 1830s energized the insurance underwriters to push for large-scale, reliable water service, built by governments at great expense but enthusiastically supported by property owners whose taxes would rise substantially because of the project. (It is the insurance companies of today who push hardest for action on global warming, which has had a significant effect on the increase in weather-related catastrophes in the 1980s and 1990s.)

Another example in Lemon's book of the transformative possibilities of public and private cooperation was the reform movement in Chicago at the turn of the century, which, while upper middle class in its composition and policy orientation, was nevertheless a creative response to socioeconomic stress in a city with a population that had exploded from 500,000 to 1.7 million in the 20 years between 1880 and 1900. Similarly, the reform era in Toronto in the early 1970s produced housing and development policies so innovative that they attracted world attention. These policies emerged from a remarkable joint effort between private citizens on the one hand and open-minded city politicians and bureaucrats on the other (See "Is It Accidental or Merely Wild?" in this volume) . In fact, what Lemon shows is that many seemingly intractable problems were solved in cities during the last three centuries by a process that made both civic and economic sense. Cooperation between public and private actors underlay new kinds of development, not just more of the same kind of growth.[1]

Contrast Lemon's perspective, now, with that of Jacobs. It is she who has given new insight into how urban creativity thrives best when one activity or physical form (or species) does not overwhelm the others. Small-scale physical diversity in the city allows all sorts of different enterprises, cultures, social habits and organizations to interact with and to feed off each other. The unexpected meetings produce the inventiveness that characterizes evolution – not just economic evolution, but also social and cultural evolution, since they are all related. This kind of growth is not more of the same thing; it is new kinds of processes and new levels of relationships. This is, in fact, how nature spins its evolutionary web. Jacobs' argument is that the very unpredictability of this creative process makes it unplannable, by governments or anyone else. One cannot plan to be creative; one needs to allow the chemistry of interaction among diverse people and businesses to work its magic. Governments can protect the test tube, which is the physically diverse city district, but when they try to manipulate the outcome of the experiment, they produce deadly urban smears with stagnant and inefficient economies and feeble social networks.

This point of view pushes Lemon's buttons. He believes that leaving the city's destiny, as the liberal ideology does, to "private property, and therefore the private corporation, has hindered the public corporation, and so the sense of commitment to the common public life." In many places Lemon clearly equates sensible urban growth with public policy to direct that growth, while Jacobs is impatient with unimaginative and insensitive public policies that she feels have ruined many city districts.

Thus *Liberal Dreams and Nature's Limits* raises issues of the nature of urban development, and also of the relations between public and private action, and of our gut ideological feelings toward those forms of action. Lemon's preference for public solutions to the economic stagnation and environmental destructiveness of modern cities gets translated, at the end of his book, into a scenario for leaders deciding to implement a program. The headings give the flavour of this program: income distribution, full employment, non-profits, election of politicians by proportional representation, debts and taxes, technology as servant, reused resources, and living together. Notice, first, that these initiatives come from above, not from a mobilized citizenry. Second, this is largely an economic program, with actual references to cities only under the last two headings: he recommends more recycling of "derelict buildings" and the mixing of classes and ethnic groups in city neighbourhoods.

These proposals are a wish list that reflects Lemon's pessimism about the possibilities for real change. I share his jaundiced view of computer technology, and his conviction that material "progress," in the sense of producing ever more goods, is nearing its end. But nature does not put limits on human transformation. Getting unstuck involves first redefining what we mean by urban growth, urban development and even human progress. We are advancing, as the Beatles said, in the wrong direction. The question is not whether we need more or less growth; the question is, What is real growth? As Lemon suspects, it is not more suburbs, office buildings, highways and sports arenas. Our current forms of so-called development increase income inequality, making life more difficult for women and the poor; draw down vast amounts of natural capital (soil, non-renewable resources, water, air); destroy healthy neighbourhoods; and – by forcing us to cocoon inside our own structures and images – deaden us to the kind of evolution Darwin showed happens in the rest of the natural world.

Now is the time to imagine human settlements whose size, physical design, and social and economic processes are ecologically sustainable and supportive of the evolution of the human species. That is, rather than surrounding ourselves with artifacts of our own mental processes (such as indoor shopping centres), we need an urban environment that surprises and excites us with possibilities, one that includes the rest of nature in as many forms as possible. We are insecure about the unexpected, but it is the unexpected that inspires us to transcend ourselves. Authentic urban growth and development, then, can be defined as synonymous with our development as a species.

The seedlings of such a transformative urban forest are appearing here and there, in cities described by Lemon. In New York's Bronx, whose rubble-strewn vacant lots and distressing crime figures are familiar to most of us, a single family, aided over the years by a growing network of neighbours and community agencies, has turned a run-down residential street into a thriving neighbourhood, with owner-built housing, community gardens, reoccupied storefronts, and gainfully employed high school dropouts. (See "The Politics and Ecology of a Healthy City" in this volume.) Here is a level of cooperation and love beyond the ken of the best-financed social service bureaucracy. In Chicago, some of the poorest neighbourhoods are regaining control over their food supply by growing much of their food in urban farming projects, which spin off into better health from better nutrition, lower crime rates, and greater social cohesion and political competence. Rooftop gardens also keep buildings warmer in the winter and cooler in the summer, which shows there is ecological sense to these initiatives as well.

Before the old city of Toronto was senselessly amalgamated with its surrounding municipalities, it had started supporting similar projects, realizing that their many aspects – social, ecological, economic – reinforced each other. These were conscious attempts to tap into the city's own multi-dimensionality. These experiments hold more promise for the city, and for humanity's evolution, than all the subdivisions, shopping centres and office towers in the world.

There is a growing literature on what sustainable cities could look like, led by authors such as Mark Roseland and Richard Register in North America. Significantly, however, there are many more examples of sensibly designed communities in Europe than in North America. The difference can be explained by the fact that in Europe local residents themselves design and manage, sometimes even construct, the infrastructure and buildings. This public participation contrasts with North America, where professional planners and architects design and write about such communities, but little sustained public action occurs. Getting unstuck at this level involves rethinking what we mean by public and private action.

To contrast faith in public action with faith in private enterprise is simplistic and in fact illustrates how we have become confused by terms. The first thing to notice is that the private economic sector is so heterogeneous that it is meaningless to refer to it as a single entity. Variety stores and AT&T are thrown into the same category. Large-scale corporations, in fact, have far more in common with government than with a local restaurant or with Stan's Hardware: these corporations and their personnel depend upon and have close-knit ties with the public sector. The variety of political and economic preferences across the private sector, in other words, is too diverse to lump them all together. The scale of firms and their location explain these preferences far better.

Second, the distinction between public and private has been muddied. John Ralston Saul has argued that corporate culture – made up of public, semi-public and private corporations and their executives – has intentionally mixed public and private interests at the institutional level. The well-being of corporations, which must be chartered in any case by the government, depends on direct and indirect subsidies, physical and electronic infrastructure, and regulatory commissions, all provided by public authorities. In this setting, argues Saul, what is called public policy has become defined as the compromise among competing private interests, a very meagre definition of the public good indeed. In fact, there is a tendency to reduce the relationships between public and private to the abstract political and legal contexts of economic interests, especially as they relate to property. Many writings on political philosophy come close to equating "public" with the political. Now, clearly we are not simply abstract political and economic beings, but also social, familial, sexual, emotional, moral and spiritual beings, and all these sides to us have public and private faces.

Furthermore, all this behaviour – all of it – must happen *somewhere*.

Our loss of bearings about what is public and what is private relates to our loss of a sense of place, which used to be an inherent part of defining the public and private parts of our lives. In particular, we live in cities with buildings and places that physically express the fact that some of our activities are private and some public. In other words, what happens is connected to where it happens, and those of us writing about cities need to be aware of this. When we are face to face with others in a particular place, the meaning of what is public and what is private is usually pretty clear. But consider that in the North American city such so-called public contact is most likely to occur in large buildings or concert halls, at our work, in a shopping centre or in a subway station – places shaped by their functions. Between buildings there are only roads, which are the preserve of whizzing cars and trucks. In such an environment, casual, spontaneous, face-to-face public contact is driven to the margins, although it does still occur sporadically just outside our dwelling units, to the extent that there is any sidewalk or street life in our residential neighbourhoods.

This fact has serious consequences for politics, which for

most of us now consists of exposure to media output, not of face-to-face discussion in public about issues of collective concern. Since most of our places, both public and private, are programmed, the media have by default become the principal arena for public discourse, an arena that renders that discourse similarly programmed, vicarious and, of course, placeless. "Public" in such a context means what Hannah Arendt has called "that curious hybrid realm where private interests assume public significance." Real, physical public interaction has all but disappeared because of the way we build, a fact Lemon obliquely acknowledges in his plea that "leaders" move "to modest-income areas in central cities. Imagine leaders and other citizens holding street dances, together tending community gardens, and riding bicycles."

What Lemon is groping for here, I'm almost certain, is the loss of community. Modern usage of the term notwithstanding, an authentic community has territory as a necessary (but not sufficient) condition. The close-grained mixture of classes, ethnic groups and land uses in Toronto's downtown so lovingly described by Lemon is, however, not just a geographical or demographic phenomenon. It forms a basis for all sorts of face-to-face interactions that are casual, unplanned and momentary, that build up over the years into more formal groupings, into affection for places and people, into wisdom about how the place and people fit together, and into mutual care and support among residents and business people. This is community, grounded in a unique place and creating a public good that transcends the sum of competing private economic interests, a public good that reflects our social, spiritual and artistic dimensions.

Wendell Berry argues that our confusion over the distinction between public and private stems from the fact that there are no longer any physical communities to mediate that distinction. As a result:

> Private life and public life, without the disciplines of community interest, necessarily gravitate towards competition and exploitation. As private life casts off all community restraints in the interest of economic exploitation or ambition or self-realization or whatever, the communal supports of public life also and by the same stroke are undercut, and public life becomes simply the arena of unrestrained private ambition and greed.

Many would categorize Berry as a communitarian, and liberalism has been set against communitarian ideals, not just by Lemon. Despite their strong differences, these philosophies have an important goal in common – defining what should be of public concern. Community gives moral substance and physical expression to the relationship between public and private, and in the intelligent development of that relationship lies promise for vital cities in particular and for human evolution in general. In spite of its diverse and contradictory elements, the liberal persuasion at its best is also a grand attempt to make sense of the dynamic between public and private rights and responsibilities within a context of self-government. When that attempt was started, during the upheavals of the 17th century, when formerly legitimate statist institutions were in question, one's physical, daily connection to a local community (however undemocratic) was still taken for granted. Cut from these territorial moorings, liberals and their dreams have drifted into dangerously abstract waters. In the interim, large-scale states with huge bureaucracies – enemies to both liberals and communitarians – have arisen to replace authentic politics of any kind.

Warren Magnusson challenges the view that nation states are sensibly scaled political units (although they vary in size, even the largest are too small for some policy problems), and reminds us that the locality remains the appropriate scale for meaningful politics. His point is that such politics must be grounded in everyday local experience. In a similar vein, I see the magnificent array of figures and facts assembled by Lemon in *Liberal Dreams and Nature's Limits* floating disconcertingly above the cities they are used to describe, disconnected from the personal experience of place for which there is no substitute. The statistics, like the terms public and private, are too abstract for the understanding of our attachments to streets and buildings and people. As an example of how such attachments are given a sensible material context, consider the front porch on city streets: it defines a clear transition zone between public and private space, and porch-sitters enjoy occupying that zone – it is comfortable both for the occupants and for passers-by. In this way community ties are defined and given life in physically defined places.

I have devoted so much space on this issue of public versus private action because Lemon's book is a good example of why we are collectively immobile. It exemplifies the view that we are approaching physical limits, in terms of dirtying our own nest, yet little is done about it. This is because, as his proposals illustrate, we still accept the old definitions. The key to the link between nature's limits and liberal dreams is public participation, which is also the community link between public and private. The technology on which nature has set limits is in fact non-participatory, thrives on passive consumerism and the cult of the expert, and has done much to destroy the local social vitality that underpins community. Local communities, especially within cities, need not be conservative; the conservatism of some stagnant communities comes from their isolation and homogeneity, not from people's affection for locality and each other.

Lemon, who participates fully in his own urban neighbourhood, here and there informs his book with accumulated wisdom from that part of his life. He admires Toronto's pragmatism and piecemeal growth (another favourite idea of

Jacobs'!). The vitality of Toronto's local politics, and of local politics around the world, contrasts sharply with narrow-minded parochialism of provincial and federal governments. Initiatives first tried in small urban neighbourhoods are shared with other neighbourhoods in the same city and across the world before any other level of government ever gets wind of the idea. Perhaps these initiatives, based on an awareness of place, can lead us past our dead-end fixation on public versus private, liberalism versus communitarianism, and cities versus nature.

NOTE

1. By contrast, contemporary urban development is characterized by so-called public-private partnerships that often become economic fiascos and make governments look foolish, as they negotiate deals that subsidize private companies and create sinkholes for tax revenues. Such forms of cooperation are clearly not a panacea.

BIBLIOGRAPHY

Hannah Arendt. *The Human Condition* (Chicago: University of Chicago Press, 1958).

Wendell Berry, *Sex, Economy, Freedom, and Community* (New York: Pantheon Books, 1993).

Jane Jacobs. *The Death and Life of Great American Cities* (New York: Random House, 1961).

James Lemon. *Liberal Dreams and Nature's Limits* (Toronto: Oxford University Press, 1996).

Warren Magnusson. *The Search for Political Space* (Toronto: University of Toronto Press, 1996).

Richard Register, *Ecocities: Building Cities in Balance with Nature* (Berkeley, CA: Berkeley Hills Books, 2002)

John Ralston Saul. *The Unconscious Civilization* (Toronto: House of Anansi Press, 1996).

Domestication as Pavement

I

I am a child of the city. Not only have I lived in large cities all my life; I have made them a central focus of my teaching and writing.

Cities are fascinating. At their best, they are celebrations of human diversity, ingenuity, and creativity. We can learn much about ourselves by exploring our own built environments and our behaviour in them.

Recently, however, it has become painfully clear that the way in which we have been building our cities is making a mess of our ecosystems, to say nothing about these cities' undesirable effects on ourselves. Part of the problem – and the reasoning here becomes a bit circular and difficult to disentangle – is that we seem to be constructing environments for ourselves that desensitize us to our surroundings, both natural and manmade.

Most of us spend a large portion of our days cocooned in buildings, or in cars, or inside some other human construction. We have organized our lives so that we experience only excruciatingly tiny fragments of non-human nature, and those fragments have themselves been manipulated and shaped by us – landscaping, house plants and cut flowers, zoos, and species imported by us from other continents (such as sparrows, apple trees, and dandelions). We are, as Jerry Mander has said, literally living inside our own heads.

In such an environment, we forget that we are part of an ecosystem, a community of plants and animals and microorganisms, that our cities are inextricably linked to their surrounding country, that humans and even concrete all came from the soil. Until we start seeing ourselves as part of this larger community, studies of the city, however fascinating, will be incomplete. We shall all continue to be children of the city, because we shall never have grown up.

Growing up, as John Livingston makes clear in *Rogue Primate*, must involve intelligent bonding with the rest of the natural world. Humans sought to even out the feast or famine cycles of hunter gatherer existence by domesticating plants, other animals, and ourselves. In doing so, Livingston argues, we lost our unselfconscious belongingness to the rest of nature and thus the ability to become mature adults.

Livingston's definition of domestication relies mostly on descriptions of animals that we – the only evolved domesticates – have ourselves domesticated:

> The domesticated mammal . . . is docile, tractable, predictable, and controllable. Initially it may be smaller, but may become somewhat larger than its wild antecedents. It grows very rapidly, but even into maturity retains many infantile characteristics both physical and behavioural. It is dependent on us. It is sexually precocious and promiscuous. There may be great variety in appearance, but behavioural individuality is low. There is pronounced reduction in sensory acuity and the ability to communicate both intra- and interspecifically. Dogs excepted, social behaviour is much simplified and low in subtlety. Dogs excepted again, there is no attachment to physical place, and no awareness of social place. No domesticate has an ecologic place. (27)

Humans have most of these characteristics, with a couple of exceptions. One is that our intra-species communication and social behaviour are quite complex. The second is that our dependency is on other humans, and on technology.

Technology becomes an important theme in Livingston's book, and figures in his tentative answer to the obvious question, why did we enter domesticity on our own? He suspects it had something to do with fire. Many animals used fire, but humans were the first to control it. The method of control became a peculiarly human technology, he argues, because it depended on "storable, retrievable, transmissable technique" (35).

Thus, while the animals we domesticated depend on us, their owners, *our* dependence is on technology. Our culture became a culture of how-to-do-it – at first more sophisticated weapons and tools, then shelters and agriculture, and, most important, social organization for the purpose of control. That is, other animals have cultures; they also have technologies. Only humans have a culture dependent on technology, one that includes the storing, retrieving, and transmitting of technique itself. This technology-dependent culture seems to be at the core of what he calls domestication.

The opening thesis of *Rogue Primate* contains some other elements. One is that the control of fire began our separation from the rest of the natural world: Livingston paints the image of early humans huddled around a fire at night, with a growing apprehension about the Otherness at their backs. Control of fire also seems connected to our willingness to tolerate what he calls unnatural crowding, which is probably the most intolerable situation for any wild animal. Furthermore, the culture of technology, as the defining feature of our domestication, became a substitute for the direct experience of nature – a prosthesis.

Devoting our lives to How inescapably condemns us to vicarious experience. This perspective provides a number of insights into humanity's current problems. I shall return to this point later.

The book contains a convincing demonstration that it is the domestic nature of our culture which is at the root of our trashing of the Earth's ecosystem. This demonstration includes documentation of the awesome impact of the introduction of plants, animals, and diseases into new habitats; and determined, ruthless extirpation of certain species, such as large flightless birds, the woolly mammoth, and the passenger pigeon. Livingston then shows how, starting in the 16th century or so, Europeans exported not only diseases and domesticated animals, but also their ideologies. As in the case of physical diseases, European ideologies met with no immune system. (One does wonder about Europeans' immunity from ideologies in the lands that were colonized.)

Local cultures had developed ideologies that in some way reflected their immediate environment. What Livingston calls the exotic ideology saw and sees itself as neutral and universally applicable. It grew from the conviction that humans are superior organisms, separate from nature but able to study and control it for human purposes. Those purposes are narrowly defined in this ideology as growth in production and in material well being. The dominance of this ideology over local cultures seems to ensure, not the destruction of the biosphere, which has survived more cosmic catastrophes, but at least the systematic removal of life-support for the human species.

Undergirding the exotic ideology, writes Livingston, is the assumption that the natural world is a system of competitive struggle for survival. He spends the better part of a chapter gleefully tearing this assumption to shreds, repeating the admonition that we see what we want to: Believing is seeing.

We are unable to see what is right in front of us, because we have set ourselves apart. To address this, the book shifts gears, as the writer becomes absorbed in an attempt to put into words the subtleties and different levels of wild creatures' belongings – to a place, to a group within their species, to other species, and to the biosphere. As he explores the many selves of animals, he shows that our species' narrow fixation on the "problem" of the individual in society is the product of an immature and stunted psyche peculiar to a domesticate. Wild animals seem to move gracefully from awareness of self as an individual, to self-as-group, to self-as-multispecies-community, where appropriate. The instinct to bond with nature is pounded out of us sometime between seven and twelve, when we are most curious to explore it and to connect with other species. It is in this sense that domesticated humans never grow up. Ironically, many famous scientists, from Charles Darwin to T.H. Huxley, recount how childhood experiences with nature created their initial interest in science; and mystical connection to nature is frankly admitted to by many contemporary scientists.

Livingston's position is that while such a bond is essential to adulthood, the human child's biological impetus to seek it is denied by domestication's socialization process and especially by what we call education. Human domesticates start training their children at an early age to use technology as the mediator between themselves and the rest of the natural world, like an artificial link. In the deepest sense, the "fully trained" human literally becomes technique. If education does go on at our institutions of teaching, then it is education of the narrowest sort, internalizing the fiction that we can only comprehend the world through our prosthetic culture.

This socialization (conditioning?) produces what Livingston calls zero-order humanism, a subconscious, unchallenged "imperative" that human well-being and human goals always come first. Anticipating the retort that every animal behaves this way, he continues:

> ... [A]ll living beings use themselves as points of reference in dealing with the world ... However ... (they) do not insist on being solo acts. They are in the centre of their individual spotlights, but if they are indeed possessed additionally of community and biospheric self-consciousness ... then they share their self-focused universes with all of those who comprise them ... The ostrich runs ... at the approach of a lion. So would you or I. It seems doubtful that the ostrich would consider it wrong ... to be captured and consumed by lions. But we would. Humans are too important to be wasted as cat food. (137-8)

The human enterprise, as defined in *Rogue Primate*, is the domestication of the planet. Since we have been trained since early childhood to be a part of this enterprise, we remain oblivious to the warped nature of our tools for this enterprise, tools that are the centrepiece of our culture: science and technology. For instance, we are blind to the hideous sufferings we impose on other creatures for purposes of our fashion industry, for our entertainment, and for research into human health – or, rather, into sickness.

This brings Livingston to the subject of animal rights. While his sympathies are clear, he dislikes the term, since it puts other species into a framework of human laws. Rights only need to be invented when there are human institutions around to violate them. More specifically, domestication requires the exercise of power-over. This is normally thought of as some humans telling others what to do; however, cultures also have, in varying degrees, internalized rules of behaviour telling us all what to do on the explicit assumption that we are stupid or too evil just to be. These internalized rules are patterned into organizations and institutions such as schools, the state, religions, and the market system.

Rights are thus an artifact of domestication, which means that there is no such thing as a natural right. As instruments of human legal and political systems, rights are irrelevant to "beings who are free, wild, and whole." (196) This irrelevance stems from the fact that such beings experience none of the tension between individual and collectivity that causes such angst for humans. Rights can only be a concern of an immature being which has not yet learned to live in harmony with others of its species or, for that matter, with the rest of the world.

A debate over the granting of rights to animals, or to plants, or to anything else presupposes someone or something that does the granting. The whole concept of rights is exposed as part of the pathology of hierarchy and dependence that characterizes domestication. Rights are an artificial substitute (a prosthesis) for Rightness, which is a natural feature of the interdependent mutualism found in the wild.

Some humans are quite sensitive to this truth. Saul Alinsky devoted his life to rightness – helping people realize that they are perfectly capable of making their own, intelligent decisions about their collective affairs, at the local level. He had nothing but disgust for the quintessentially domestic process of certain people (however chosen) making decisions for others, and quoted with approval Mr. Dooley's acerbic little speech:

> Don't ask f'r rights. Take thim. An' don't let anny wan give thim to ye. A right that is handed to ye fer nawthin has somethin the mather with it. It's more thin likely it's only a wrong turned inside out. (124)

Rogue Primate constitutes one of the most unflattering assessments of our species and of its noxious effects on this planet that I have ever seen.

II

So where do we go from here? The last chapter of the book does not contain specific proposals or programs, which frequently are strategies to avoid real change. But Livingston is eager to awaken us to the warped nature of this culture's received wisdom. There is an appalling enormity to such a project. Livingston has been talking about millennia of conditioning. Nevertheless, he concludes,

> The experience of wildness. Like its close kin which we call freedom, wildness is perceptible only its absence. Both are forever paradoxical. Percy Shelley saw freedom as "sweet bondage." We may see wildness similarly: a state of being in which one is an autonomous organism, yet bonded and subsidiary to the greater whole . . . [O]ne is at once the end and the means, a unique expression and totality. (196)

In declaring that this state is still accessible to us, and that it is essential to redressing the crisis we now find ourselves in, Livingston surely opens himself to be easily and lightly dismissed. I can just hear people saying, "Sure, all we have to do is feel at one with nature and all our problems will be solved. Give me a break."

Anyone who has followed the argument this far can see how such a response trivializes Livingston's argument. What he has done is take some specific but absolutely central contemporary issues – ecosystem breakdown, authoritarianism and lock-step followership, rights of the individual, humans' mistreatment of other humans, the debates over free market competition – and shown how they fit together, by linking them to domestication. This is such a fundamental and ancient part of our character, though, that one could still see reason to protest. From a heuristic point of view, domestication is such a cosmic variable that it could be used to explain everything; and in explaining everything, it explains nothing. From a practical point of view, the conditioning of domestication is by now ineradicable in humans.

There is no completely satisfactory answer to these objections, since Livingston is redefining heuristics and practicality as well. He is writing in the tradition of philosophers and mystics who are challenging us to raise our awareness of our assumptions – of our presumptions – and of patterns we treat as inevitable in our daily lives. It is the very experience of a new awareness, a new understanding, that is sufficient to change our behaviour. Once that awareness and understanding is experienced, in other words, the direction of one's life becomes changed. This need not be a so-called peak experience, but it usually involves a sudden, almost instantaneous comprehension of how many things are connected.

One example of this process is given by Doug Elliott, a

North Carolina naturalist. As a boy, he had been fascinated by snakes, and as he grew up he became more and more knowledgeable about them, learning how to handle even poisonous ones. When he was in his twenties, he came across a water moccasin.

> I quickly pinned the large water moccasin's head beneath the rim of my net. Then I grasped it firmly behind the head and picked it up . . . The snake writhed and twisted as it tried to free its head and neck. I restrained the middle of its body and allowed the tail to flail about . . . It would have bitten me if it could have, but my technique, rehearsed hundreds of times in my mind and with harmless snakes, was flawless. Its jaws gaped and its fangs came forward unsheathed and drops of venom, clear and glistening, dripped from their needle-sharp tips. I definitely had control of this snake. But what kind of control did I have? I had captured this snake with the skill of a professional, something I had dreamed about since childhood, but there was something less than satisfying about it. Is this the way I wanted to relate to nature and to the world in general?

Sometimes afterward, he was leading a group of backpackers through the southern Appalachians, when they came across a magnificent timber rattler. Elliott borrowed a hiking stick:

> I gently lifted the snake and moved it a few feet onto more open ground. Its head was raised and curled over the forward part of its body. The snake looked at (us) . . . with a timeless, emotionless self-assurance . . . It did not rattle nor did it coil up defensively. Just its tongue flickered out, probing the highly charged atmosphere for a molecule of meaning. Could it sense that we meant no harm? . . . The posterior part of the snake was stretched out, its tail pointing in my direction. I slowly reached out and touched the tail. The snake did not show any sign of irritation. Gazing into its eyes, I slipped my hand underneath the tail and lifted gently . . . Sometimes, during the normal course of a snake's life, part of its rattle breaks off. This does not hurt the snake . . . I carefully grasped it between my fingers and gently twisted the last four segments off the rattle. I let the tail go and looked at the section of rattle I now had in my possession. I could hardly believe what I had done. It was certainly nothing I had planned

to do. *Planning* . . . would be even more foolish than actually doing it. (16-17, 22-23) (Emphasis in original.)

When there is a gap between feeling and action, that is planning – carefully weighing one's values, hopes, and fears against some future goal. Action then becomes devoid of spontaneity, of life, and indeed, of effectiveness.

This is why it is so important to understand Livingston's avoidance of a prescribed course of treatment. Such a prescription could only flow from our conditioning, from our past. The act of seeking prescriptions, of formulating public policy, in fact, is the essence of domestication – the culture of How. And, paradoxically, it is the essence of inaction: "We pursue the ideal because it doesn't demand immediate action; the ideal is an accepted and respected postponement," writes the philosopher J. Krishnamurti. (160)

As an antidote to planning, and to the seeking of ideals, Livingston and Krishnamurti are asking us to see, without artifice, without intervening thoughts and images. For Krishnamurti, it becomes an act of love. For Livingston, it is re-experiencing wildness. Whatever one calls it, this certainty of wisdom is immensely powerful, and the only source of real change, since it is free of the past.

This is indeed a different mode of thinking; in fact, thought itself generally gets in the way. Many Westerners object that in talking about some profound change in an individual's consciousness, mystics are navel-gazing, completely ignoring society at large. That is not the case. This objection is an example of the domesticate's mistaken view of social reality, that the individual and the group are separate entities. We cannot conceive of an individual who is autonomous but simultaneously an integral part of a community. To a domesticate, this is an oxymoron. To a wild being, it is the definition of existence.

It is important to make the distinction, in this context, between human social organization, which is hierarchical, and the social organization of other species, which tends to be mutualistic and cooperative. As a naturalist, Livingston has some interesting perspectives on our mistaken perceptions of hierarchy among other animals. But he is also eloquent on the awful ways humans treat each other.

Domestication involves control. It is a technique of controlling the behaviour of other species, or of other humans. In the case of humans we use a conceptual trick that Livingston calls pseudo-speciation – identifying other humans as another species, which allows us to treat them as unconcernedly and brutally as we do other animals. Domestication's control requires that the controller see himself as separate from other humans and other species, whoever is being controlled. While in the long run this can only be a false perception, humans maintain the fiction in order to perpetuate hierarchy.

*

Rogue Primate suggests that withdrawal from wildness started everything. This does not mean that our only focus should now be on individuals' bonding with nature. We withdrew from wildness socially as well as individually – it was two sides to the same coin. Besides, Livingston maintains that we shall always need some sort of cultural prosthesis, some sort of ideology, since we cannot turn back the clock. But the essential point is that he is convinced that participatory compliance and reciprocity evolved naturally among all animals, including humans. Though he dismisses the term social ecology at one point, he would have no problem with social ecologist's advocacy of decentralized, cooperative societies.

It makes perfect sense to ask, of course, Are not cooperative communities an ideal? Are they not here being treated as a goal or a solution to strive for and thereby simply another expression of the old ways of thinking? My answer would be yes, cooperative communities could be an ideal, or a policy, if we chose to see them as such. And this would ensure our failure to create them.

Chasing after a future ideal is one of the symptoms of the prosthetic ideology, which serves to keep us from direct experiences such as participatory compliance. The ideology requires, first, that the idea is congruent with the human enterprise of domestication, and, second, that a method is developed to reach the goal, thus separating ends from means.

Congruence with domestication means that the ideal needs to be rationalized in such a way that the basic question of rightness is ignored and replaced by the question of whether the goal is good for humans. Thus, Neil Evernden has shown that most environmentalists have resorted to arguing that the destruction of ecosystems by humans is stupid, not that it is wrong. Confronting the morality of environmental damage puts zero-order humanism itself in question, which is a no-no. By the same token, as long as we are under the spell of the prosthetic ideology, we cannot justify cooperative social behaviour unless we can prove that it benefits our domesticated institutions and values.

Second, the ideal of cooperative communities, since it is something to be achieved in the future, needs to be provided by our how-to-do-it culture with a technique for reaching it. There is, of course, no method to spontaneous cooperation, but our ideology prevents us from seeing this. And because the ideal is simply a projection from the past, our technologies are just helping us do more and more of the same thing, only faster. The Information Highway is a grand example of this phenomenon; we are now able to find out about more cooperative communities, about research into cooperation and its benefits, and about philosophical discussions on its merits, without ever having to do anything.

If we understand instinctively that cooperation is right, then we don't have to prove that it's smart, or to develop a method to reach it. Cooperation instantly becomes a pattern of behaviour, from which many different kinds of collective action will emerge, unplanned, but somehow right.

We do not trust that it will be right, though. Millennia of domestication have made domination and dependence the air we breathe. This may be a difficult pill to swallow for those of us who live in what are termed representative democracies and who feel they provide us with a reasonable sense of freedom of speech and thought, despite their weaknesses. Sense of freedom, however, is like language: the other person always speaks with an accent. It is easy for us to see the constraints imposed by other cultures, not so easy to see our own constraints, our structures of dominance and dependence.

Consider, for a moment, that in our so-called democratic society, we depend on a huge entertainment industry to keep us entertained. We depend on a sickness industry if we start feeling sick. We depend on an enormous production and distribution system to feed us agricultural products of doubtful nutritional value. We depend on a massive system of compulsory schooling, not only on what to teach us, but how to teach us. In general, we do not create our own work, but look to large institutions to provide us with "jobs." And, we look to others to plan and build our cities. These are structures reeking of domination and dependence (the two are inseparable). No wonder that when spontaneous cooperation starts to accomplish something, we are distrustful. The chains of domestication are practically invisible, but they are tight. It is especially galling because we have forged them ourselves.

The invisibility of these chains has been dealt with, appropriately enough, by writers on the psychology of separation from nature. Morris Berman, Erich Fromm, Paul Shepard, and Theodore Roszak have all explored this separation as a form of collective madness, a madness that explicitly includes feelings of separation of humans from each other, as well as from the rest of the natural world. Because it is collective, our madness appears to us as normal. Our amnesia is so complete that we treat feelings of belonging and of total awareness with skepticism and derision, since they are aberrations from our domesticated mindset.

Collective madness, however, must be collectively sustained (a perverse form of cooperation!) because we still have enough access to wildness, to use Livingston's term, for that wildness to penetrate our consciousness from time to time. We have set up elaborate structures that serve to convince us that this is indeed an aberrant experience. One recently organized and extremely effective structure serving this purpose is compulsory schooling, which in a thousand ways teaches us to remain childish, dependent, and incomplete – in short, domesticated. I have learned, for instance, that the mark of a competent university teacher is to be knowledgeable with respect to a certain amount of accumulated information about the past and to transfer effectively some of that information to the passive minds of students. My competence, in other words, depends on storable information, and my students' "education" depends on me. In such a system, we all remain deprived of our wholeness. Instead, like Mr. Dooley and his rights, we should be suspicious of an education that is "given" to us.

Many other institutions of domestication – the market system, the legal system, the military-industrial complex – are in place and seemingly unassailable. They are not abstract, nor are their consequences, but they are intangible enough that it is difficult to be aware of how their power is supported by our habitual behaviour.

One structure, I believe, though, is more tangible and lends itself to direct personal action and therefore to change: urban development. Actually, the term "urban development" already puts it out of reach; perhaps we should just call it our buildings and our settlements.

Parts of some cities, I submit, can be considered wild because they have not been planned. (I exclude the senseless unplanned sprawl that surrounds most of our cities.) In these places, peoples' energy and intelligence have produced, over time, an intricate web of land uses and interpersonal relationships that are definitely subversive of the domesticated ideology. These relationships cannot be neatly categorized as economic, as social, or as political (often they are all three), but they *are* interdependent, mutual, and caring. Such places reflect a mature trust in the harmony of participatory collective enterprise, a letting go of the childish dream of domination. They also tend to have unexpectedly large amounts of greenery, both planned and unplanned. In such places there is a chance to grow up, indeed. Some are described in other essays in this book.

User-designed settlements activate the domesticate's fear of wildness – anarchy, we call it. "Nothing will work," we say. "Nothing will fit together." Wildness seems chaotic to us because our underdeveloped senses do not discern its patterns and rhythms, which are not meaningful in abstract, only in experience.

Yet part of us still resonates with those patterns, still can feel what the architect Christopher Alexander calls the Quality Without a Name, a quality that suffuses places and spaces that feel just right to us, though we don't know why. They are alive, and timeless, and they fit unselfconsciously into nature.

Thank goodness the Quality has no name. If we were able to bring it into our language, we would kill it with labels. Like wildness.

In the city, wildness happens when we aren't looking, when we aren't planning. Cities that do not harm the ecosystem cannot be reached by a policy or a program; they are far more accessible when we stop doing the things we already do. Then, in the cracks in the pavement, dandelions and plantain unceremoniously push their way into our consciousness.

The pavement is a metaphor for the domesticated mind, as well. Most of the time it is cluttered with laws, rules, and plans, with domesticated chattering. But, occasionally, exhausted, it leaves a space and the truth creeps in. John Livingston has crept through one of those spaces.

BIBLIOGRAPHY

Christopher Alexander. *The Timeless Way of Building* (New York: Oxford University Press, 1979).

Saul Alinsky. *Rules for Radicals: A Pragmatic Primer for Realistic Radicals* (New York: Random House, 1971).

Neil Evernden. *The Natural Alien: Humankind and Environment* (Toronto: University of Toronto Press, 1985).

Doug Elliott. *Wildwoods Wisdom: Encounters with the Natural World* (New York: Paragon Press, 1992).

J. Krishnamurti. *Commentaries on Living* (Wheaton, Ill.: The Theosophical Publishing House, 1986).

John Livingston. *Rogue Primate: An Exploration of Human Domestication* (Toronto: Key Porter, 1994).

Jerry Mander, *In the Absence of the Sacred* (San Francisco: Sierra Club Books, 1991).

What Is Sustainability?

Heidigeer in Hamilton

As recently as 2000 BC, the Mediterranean's shores were lined with oaks, beeches, pines, and cedars. The cedars of Lebanon grew straight and over 100 feet tall – until they were all cut down by the Phoenicians for trade and for their own uses. The Greeks and the Romans cleared more and more of their surrounding forests, replacing them with pastures where goats and sheep cropped vegetation so closely that nothing grew to hold the soil. By the time of Christ, the Mediterranean landscape had become the one with which we are familiar: rocky and bare hills that support hardy vines, rosemary bushes, and olive trees, but little else. The climate of the area had become hotter and drier because of the lack of foliage. For thousands of years, humans have been far more effective at trashing their environs than looking after them. So, although the idea of living sustainably has been around as long as human language has been there to express it, having the idea is a far cry from putting it into practice.

The publication of *Our Common Future*, by the World Commission on Environment and Development in 1987, put the concept of sustainable development into the forefront of philosophical controversies, policy formulation and decision making, and political maneuvering. Sustainable development, defined by the WCED as development that meets present human needs without compromising the needs of future generations, ended up influencing attempts all over the world to build sensibly, while at the same time prompting vigorous criticisms of it as an excuse for business as usual.

Ingrid Leman Stefanovic teaches philosophy at the University of Toronto. Her purpose in her book *Sustaining Our Common Future* is to show how insights derived from phenomenology, in particular the thinking of Martin Heidegger, can help us to arrive at a critical, balanced view of what sustainability really means.

Right at the start, Stefanovic declares that phenomenology's utility lies in its ability to "expose taken-for-granted assumptions, value judgments, and even cultural paradigms and language structures that condition our way of seeing the world."(xvi) Her book comes back to this theme again and again, most particularly in the context of how westerners impose their narrow concept of development and their mechanistic methods on Third World Societies. For instance, the book has an excellent section that critiques the whole notion of sustainable development indicators (SDIs) which, while they are an attempt to focus our energy on development that is ecologically friendly, are still stuck in what Stefanovic calls our calculative mindset.

She applies this critical stance to two major themes in the sustainability literature – ethics, and our sense of place – and then proceeds, quite appropriately, to intertwine those themes. Her analysis wrestles constantly with the problem of transcending western dualities such as those of the general versus the particular, of the cosmic versus the local, of rules versus spontaneity, and of aggressive manipulation of nature versus fatalism before its power. For example, there is an articulate discussion of a reconciliation between the totalitarianism of top down moral precepts with the anarchy of pure cultural relativism. As elaborated below, that discussion argues that there are some universals – not so much specific rules as ways of engaging ourselves in an ethical process.

In this context, place becomes crucial for Stefanovic. We are always in some place at some time, and that "implacement," as she calls it, helps define us as well as our ranges of ethical alternatives. Our places are inside as well as around us (rendering the term "environment" peculiarly clumsy); and they are not static but always changing. Ethics thus becomes a moral process interwoven into our daily life, in our own place, not some abstract set of guidelines for us to follow anywhere at any time, nor an arcane academic discipline relegated to specialized journals and books. From this perspective, instead of being a technical fix, sustainability turns into a graceful symbiosis between our presence in a place and ethical behaviour in that place. These are solid and important arguments.

From time to time, the theoretical points are illustrated with case studies. For example, Stefanovic writes about her experience of being hired by a multi-disciplinary research project on sustainable practices in Hamilton Harbour. Her job was to provide an integrative framework for the disparate research teams, grouped into the following categories: Human Values and Perceptions, Contaminants, Biotic Recovery, and Policy Analysis and Economics. Her method was to use a "phenomenological interview process" that involved extensive open-ended conversations with all the researchers. They were asked, for instance, to imagine describing Hamilton to a friend unfamiliar with the city and then were prompted, in this context, to identify important development issues facing the area. Key words and concepts from the interviews were tabulated and "ordered." Stefanovic argues that what emerged from this process was neither a simple addition of different research perspectives nor an attempt to force them all into one Procrustean paradigm.

Instead, she writes, "the interactions between these descriptors were mapped on a four-dimensional, computerized matrix, constituting an interdisciplinary, integrative overview of key elements of the Hamilton Harbour ecosystem" (156). The heuristic result, apparently, was to indicate that the collective perceptions of the teams had some significant blind spots, which skewed the focus of the research programme.

This last example can be used as a jumping off point to raise a few questions about some core arguments of the book. While the interview exercise may have been useful to the Hamilton research project, the particular contribution of phenomenology and its four-dimensional matrix is less than clear, at least from Stefanovic's description of them. My strongest reservations about the book revolve around this point: whether phenomenology is needed by the author to make her arguments. Maybe, as in the case of the bourgeois gentilhomme who didn't know he was speaking prose, many of us are phenomenologists without knowing it. But do we need to?

An extremely important question raised in the book is an apparent paradox concerning urban sprawl. We have known for three or four decades that North American suburban development is outrageously expensive for both governments and individuals, that its need for the automobile has produced massive amounts of deadly pollution, and that its social and political culture is somewhat suspect. This last characteristic – suburban culture – pushes many buttons, of course, because more than 50% of us now live in the suburbs. It might also be delicately mentioned that government majorities are based on suburban support, so political authority flows from inhabitants of this form of development.

Stefanovic wants to know, as do many of us, why suburban development continues unabated when it seems undesirable from so many vantage points – when it is, in fact, completely unsustainable. Her application of phenomenological thinking to the problem produces a couple of points. First, she refers to a theme in the phenomenological literature that stresses the fundamental significance to humans of a natural rootedness to home places, of the home as haven or even as primordial cradle. Suburban living is focused essentially on individual homes and on the womblike experience of the car and is therefore expressing (I would say taking to an extreme) "the most concrete reality of all: the need of a place that will nurture and protect its inhabitants from intrusion" (114).

Second, she provides what she calls a "a phenomenological reading" of a suburban community. This seems to involve identifying half a dozen spatial and psychological dimensions to the experience of suburban living, such as "the rural ideal preserved," or "privacy and enclosure." Another one is described as follows:

> Reflective of time as the present: The past is secured in architectural features reminiscent of a past era, frozen in discrete images. The future as sheer possibility and openness implies risk that is absent, overall, from this community, which projects an image of security in full presence. (163)

This "reading" exercise, she says, will help us "understand the foundations of residents' lived experiences of the places wherein they seek to dwell," basically as a guide to planners (164).

There is not a lot to grab onto with this kind of language. While Stefanovic's impressionistic description of the suburb is perceptive, and while it may give us some clues as to the actual content of suburban culture, the whole approach doesn't get us much closer to understanding the suburban paradox. One might note that suburbs have been around long enough for so many of us that we are likely to accept them as a normal model for home place and cradle. In phenomenology's terms, they are now the ground of North Americans' being. This is the landscape we know and accept, identify with, and even rejoice in – Stefanovic cites research to this effect. Sprawl's dysfunctional aspects become invisible to us. (My own biases must by now be crystal clear!)

If phenomenology is "a way of seeing what is right before us" (164), it should help us become more aware of these dysfunctions. Another of the book's major themes, on the primacy of ethics and ethics as a process rather than a set of rules, could be fruitfully applied to the suburban paradox.

The argument could be developed as follows. As mentioned above, Stefanovic argues that "ethical discernment is less a matter of intellectual construction than it is one of attunement to a particular way of being-in-place" (128). An example from my own life involves diet. When my wife and I were starting a family, we began paying more attention to what our children ate than we ever had to our own food habits. This sensitivity led us to investigate alternative books, journals, and food sources. It was a self-reinforcing process, so that little by little we were eating less meat and more organic food, as it became available. We joined a food co-op whose other members had embarked on a similar exploration. We did not become vegetarians overnight, but each decision led us to other ones, in a recursive manner. It was an ethical process, closely related to where and when we were living: a young family in rural Alberta, downtown Frankfurt, or Panama City might well have gone through the same sort of ethical process, but clearly with different results.

But we are concerned with the suburbs as a place, in 2002. They are a superb example of monoculture: the dwellings and their siting vis-à-vis each other are all pretty much the same. Our biological need to be rooted in a home place is now being met by developments devoid of any personality except the desperately evocative names given by their builders: Deer Glade, Highgate Hill, or River Run. The architecture of sprawl, in other words, is the concretization of placelessness. We have constructed places that numb us to any sensitivity to place. If these places are supposed to give us our identity, as the phenomenologists contend, then their placelessness guarantees that we have no identity – or, rather, that we have one fabricated by housing marketers.

More significantly, if thoughtful interaction with our home places is at the core of ethics, then our placelessness also makes us unethical. We are incapable of making choices that will sustain us in the biosphere.

The description Stefanovic makes of the ethical process could also offer some solace to those of us who feel defeated by the inevitability of suburban development. As a reflective, recursive, self-reinforcing enterprise, the ethical process is one that "reject[s] the usual science model which 'explains' every event by constructing it out of the forms and pieces of earlier events" (124). The science model is an example of what Stefanovic defined early in her book as calculative thinking: "In calculation, one studies, organizes, and computes explicitly given, empirical realities without pausing to inquire originatively about the essential meanings that sustain these investigations" (23). The cause and effect thinking of the scientific model directly informs most urban planning and therefore directly informs Stefanovic's continuing references to how phenomenology can help planning policy. In many places in her book, she suggests that policies for sustainable development, presumably formulated and passed into law by governments, can be improved by attention to phenomenology's epistemology.

Yet policies and planning are the clearest possible examples of calculative thinking and the antithesis of the ethical process she defines. Urban sprawl occurs because most of us are supporting it with our daily behaviour. Raising one's consciousness about sprawl and changing one's behaviour accordingly can't be planned. The process, as elucidated by the phenomenology literature, doesn't need any specific starting point or model of causal sequences. Purposeful action often follows quickly from awareness, and it is interesting to see the many different ways such awareness is arrived at. In the Toronto area, developers' assault on the Oak Ridges Moraine seems to have served this function for some suburbanites. As long as this sort of behaviour is emerging, and emerging from the heart of suburbia, sprawl is not inevitable, even though the power of developers seems to make it so sometimes.

Sustainability implies stasis. Conventionally, it's not a very exciting term and implies some kind of stolid, long-term good sense. Development, on the other hand, is normally thought of as dynamic. Stefanovic, to her credit, sharply questions conventional use of both words. She specifically links development to education, to unfolding, and to evolution. She quotes Thomas Langan approvingly in his effort to define development as something other than (not just more than) economic growth: "[T]he condition for the possibility of a healthier strategy of development is some collective advance in ecumenic wisdom" (142). Even here the language is one of conditions and strategies, but there is a crucial dimension brought in by the word "collective." Humans have always lived in communities, not just human ones, but multi-species communities, whose members are intricately interdependent. John Livingston, in his book *Rogue Primate*, has a wonderful chapter on how easily other animals move from a sense of self, to a sense of group (others of their species), to a sense of community (all life in the neighbourhood). Humans, by contrast, are stuck in a self/other duality:

> It is the wholeness of the wild animal that makes ethical constructs unnecessary – indeed probably unthinkable. Why create an abstract set of rules and guidelines when you are already doing all the right social things . . . ? Rules and guidelines are for . . . infantile, self-centred [humans]. (103)

The phenomenology literature, or rather Stefanovic's exposition of it, contains only glimpses of this dimension of human ontology. Yet the malaise of our era is connected, in part, to the disappearance of local human communities. At the street level, suburban neighbouring networks are feeble; instead, we celebrate our numerous placeless "communities of interest," made possible by the technology of electronic communication.

Nevertheless, by drawing on Stefanovic's discussion of sustainability and of phenomenology's central figure, Heidegger, we can connect the absence of community to our difficulties with living in a sensible way on the planet. Rather than seeing humans as Hobbesian bundles of needs, a perspective which is still at the core of the WCED's definition of sustainability, Heidegger says that the meaning of being can be found in the desire to care for all Being. My own less abstract interpretation of this is that we have an instinct to care for each other and to give to each other – and by extension to the rest of nature. This is the care-fulness that "makes ethical constructs unnecessary." It also builds community, and always has, at least until now. Finally, in caring and giving, we are participating and interacting in the world, a dynamic that leads to the unexpected and to the unplanned, which is the source of change and evolution.

This was a frustrating book for me to read, apart from the challenge of deciphering the phenomenological language. The author raises questions that I consider to be really crucial; but her answers, as I have suggested, are not that satisfying, which is why I have attempted to suggest some extensions to her analysis. Nonetheless, it is a wide-ranging and well-researched book dealing with important and difficult issues, upon which she sheds considerable light.

BIBLIOGRAPHY

John Livingston. *Rogue Primate: An Exploration of Human Domestication* (Toronto: Key Porter, 1994).
Ingrid Leman Stefanovic. *Safeguarding Our Common Future: Rethinking Sustainable Development* (Buffalo: State University of New York Press, 2000).

Culture and Granite

My wife and I have been to beaches at the extremities of Canada, from Long Beach on Vancouver Island, with its sweeping curve of sand driftwood, to a cold cove of clattering and infinitely smooth green stones just north of St. John's. They are compelling and magical places.

But it was the shores of eastern Lake Superior that first touched, then fascinated, then inspired, then finally entered me. They moved me to reflect on how we as humans emerged from the soil and rock and moisture of this planet. They moved me, in fact, to wonder about the quality of the relationship between this land and its people.

True, the historical context helped. We visited Superior for the first time in many years just as the hype preceding the referendum on the Charlottetown Accord to amend the Canadian Constitution was getting feverish – early fall, 1992 – and then again in September, 1993, just before that year's federal election was called, but long after the campaigning had started. Lured back in 1994, we arrived saturated with the debate over Quebec sovereignty, linked to the imminent electoral victory of the Parti Québécois.

Already weary of the rhetoric about Canada's nationhood, or lack of it, and surrounded by the physical reality of the Lake Superior landscape, I was caught off guard by a rogue thought: What passes for public issues, especially as packaged, reported, and interpreted by the media, is an irrelevant sideshow. This torrent of words and images invariably gives governments a central role in connecting the people with each other, and even with the land and its "resources." Actually, of course, governments can do nothing of the sort. We have more government than ever, and we are less than ever connected with each other or with the land, although it is clear that we are desperately in need not just of a connection, but of an intelligent one.

My fellow city dwellers and I, in particular, lead lives that make us oblivious to the physical environment, whether natural or manmade. Out city has been built for us, a desert of identical dwelling units, malls with the same chain stores, faceless industrial parks, and office buildings. These elements of our built environment, no matter how strenuously they may be presented to us as prestigious or exclusive, are numbing. They desensitize us to subtleties of texture, sound, smell, and colour that are so abundant in what we call the wilderness.

Because our exclusive communities and prestigious offices are spread out so extravagantly over the land, we spend much of our lives scurrying from one part of this ignorable cityscape to another. Our social and economic activities are broken up into separate landscapes, so that our relationships are fragmented and placeless. No central thread holds them together. It is no coincidence that we urbanites are disconnected not only from the natural and built environments, but also from each other. Much of what I experience and read suggests that fragmentation and disintegration are growing in Canada's rural areas as well.

The Lake Superior landscape jerked me out of these patterns of disconnectedness. Its impact on the psyche is powerful. There are no awesomely high peaks as in the Rockies; as broad as Lake Superior is, its waves have crossed distances less vast than the Atlantic's or Pacific's. Yet the landscape's features have such presence that comparisons of scale become meaningless. It is rugged and "big," of course; but there is an energy in the rock that – if you are open to it – takes your breath away. The hills are solid with sparkling red and gray granite, streaked with quartz. Here and there are black outcroppings of basalt, a reminder that in this place, an incomprehensibly long time ago, volcanoes erupted and lava flowed.

On the beaches, some of this rock has been ground by Superior's waves into miles of sand and dunes. In other places, the process is still under way: there are millions of smooth stones, from the size of a pea to that of a watermelon, lining the shores. In the centre of this rough, even brutal country, one can find granite, basalt, and agate stones that are exquisitely rounded and silky. Their colours are bright red and jet black, mottled gray and white, and bright green.

The granite cliffs and valleys that reach eastwards, inward from the beaches, are no less spellbinding. Ravens and crows circle the high points, and brilliant finches flicker through the birch, cedar, and maple trees. Although the vegetation is lush, the massive presence of the Precambrian Shield is everywhere. The plants just adapt, seeking out miraculous niches on smooth, rock surfaces. Luxuriant moss abounds, too; but I am especially awed by the roots of cedar trees, which behave more like ground-level branches, snaking across unbroken rock for yards, until they find a crack with some collected nutrients.

Once, on a hike through the woods, we came upon a birch and a cedar which had simply merged at the base. Below, the birch's roots headed down into the shallow, spongy soil, while the adventurous cedar's struck out for the far edge of a rock; above, they joined into a single trunk for a foot or two, cedar bark on one side, birch bark on the other, followed by the appropriate branches, needles, leaves.

Not far away, we came upon one of those innumerable holes in the ground, obviously home to some small animal – a chipmunk? A weasel? This particular entrance was lined on all four sides with thick moss, looking as carefully shaped as if it were in a suburban front yard; and attractively scattered around this moss, if you please, were brown yellow mushrooms, looking for all the world like lawn ornaments.

It is hard to imagine the complexities that knit together the elements of this ecosystem, this stone, moss, cedar, birch, raven, chipmunk, bear – yet I am as convinced of our symbiosis with them as of their symbiosis with each other.

This land doesn't grow my vegetables, I thought. It consists of an impenetrable bush of evergreen and marsh, overlaid on an awesomely old and enormous chunk of solid rock. It has never supported more than a few humans per square mile. Somehow, though, it spoke to me. It created a connection I no longer had the sense to make in my daily life.

The communication, surely, was wordless, but a translation would be roughly this: If we are not relating intelligently to each other or to the land outside our own front door (and we are not), then no government can do it for us, either with social services or with regulation – especially environmental regulation. This relating process is something we must do ourselves. It is hard to remember sometimes that governments and educational systems are nothing but reflections of the human psyches that created them. If our love, expressed individually, for fellow humans and for the Earth is so feeble, then our institutions can hardly be expected to improve on the situation.

This absence of love has tangible effects. Most of us throw out too much garbage, have neighbours whom we don't know or even get along with, rely on fruits and vegetables from California, pollute our environment in a thousand ways, and have some fears about our safety from robbery, assault, or terrorism. Moreover, many of us feel trapped in a job or relationship or lifestyle that isn't living up to our expectations.

In short, there is much to do. Given this situation, isn't it ironic that so many people can't find work? The problem is conveniently defined, rather abstractly, as unemployment. In other words, rather than defining the problem in our own lives and in our own places, we give our attention to the issue of unemployment, one of an array of national and provincial concerns framed for us by the media, whose very nature is to report on broad, general stories which are of some interest to everyone.

As a result, the media tend to direct our concern towards large-scale crises and dilemmas whose resolution lies far beyond our grasp, automatically rendering us powerless. Mass communication, to the extent that it communicates at all, creates a world hopelessly removed from the magnetic and visceral message of northern Ontario granite, which (at least to me) is direct, personal, and authentic. It is a message of linkages between time and beach pebbles, between cedar and birch, between a human and the soil. The unlinked messages of the media must be expressed in short, snappy sentences that take a minute or less to deliver on television, to the accompaniment of eye-catching images. We have to be coaxed and cajoled into a lukewarm interest in the so-called issues of an election, and for good reason: the definition of those issues is driven by a desire to win the election, not to solve actual problems.

The vacuity of public issues applies with special force to the case of Canada's nationhood and Quebec separation. Consider for a moment how we have hypnotized ourselves with language. If Quebec separates, do these words mean that several quadrillion tons of rock and soil will be removed to some other part of the planet? Of course not. The whole debate has really been about the survival of governments and their spheres of influence. It has nothing at all to do with our connection to each other or to the land that is Canada. Not to the land of picture postcards of the Rockies or of Peggy's Cove, or even of Lake Superior, but to the places where we live and work. This land has a life and identity we have almost totally ignored, clearly evident in our filthy lakes and rivers, heavy yellow air, and PCB-soaked soil. Quebec is no exception; just like the rest of Canada, its polluted ecosystem is nicely complemented by rising rates of homelessness, crime, divorce, and violence between humans. People actually starve and go homeless because they haven't enough money. Caring local cultures are dying, all over Quebec and all over the rest of Canada.

Healthy local cultures, as described by Wendell Berry, are sustained by people who care for each other and for the land beneath them. For this reason, they need have no fear of losing their language or customs. Such cultures are sensibly rooted in the energy of the locality – its rocks and soil, its water, its plants and animals, its human and geological history. This rootedness does not mean that the language and customs of a strongly caring community do not change. On the contrary, they are always evolving in response to growth from within and to alterations in their environment, following a process similar to that of plant and animal communities.

An acorn, for example, transforms itself in part by using the soil, air, and water right around it; but it also draws on a timeless and placeless organizing principle that in-forms it into an oak instead of a maple or a dragon-fly. And the first oak itself emerged from this creative mix of local chemistry and non-local patterns. Healthy cultures transform themselves in similar ways, because local people are constantly creating their culture out of their surroundings, but at the same time resonating with cultural principles from the collective unconscious of humanity. They will not "lose their identity" any more than the oak, since they are participating actively in their place, working with soil and plants and each other to grow and develop. In most of North America, on the other hand, we have stopped participating in this way and willingly turned ourselves into receptacles for an external corporate monoculture, whose mission has been to persuade local people that they are defined culturally as a bundle of needs that can only be satisfied from without. We know we have lost something; yet our governments, for their own purposes, have for years exploited our lack of connection to each other and to our locales. Once again, Quebec's government is no exception.

It is therefore clear that to believe a government that says it will protect a local culture is to ignore reality. Governments everywhere facilitate the invasion of mass culture into the far corners of their country, be it with the mass media, retail chains, big factories, or mega-projects.

If governments are incapable of dealing with our social torpor and the environmental crisis, then who can? I am reluctant to put my trust in private enterprise, a course often suggested in response to criticisms such as mine of governments' abilities and agendas. Many (though by no means all) business firms stand ready to tolerate pollution and homelessness as unfortunate by-products of job creation and profit seeking. A number of our largest corporations are in the business of homogenizing our culture with film, newspapers and magazines, TV networks, computer technology, and Disneyscapes. No, if we personally do not care for each other or for our land, then no amount of national identity, no amount of government money, no amount of industrial growth will save us from the consequences now looming before us.

Humanity's cultural evolution depends on rediscovering our inescapable physical connection to our local places, on becoming aware of our interior beaches. Contrary to what MacDisney tells us, this process of rediscovering the local place is a lot of fun. Indeed, the greatest joys can be found there. That is what the rocks of Lake Superior told me.

BIBLIOGRAPHY

Wendell Berry. "The Work of Local Culture," in Berry, *What Are People?* (San Francisco: North Point Press, 1990), 153-69.

In North America, we are generally desensitized to our surroundings, whether they are buildings or forests. This lack of awareness makes it easier to accept the fact that cities, towns, and suburbs are all built for us, not by us. It also makes sensible urban planning or policy difficult. The results have not been pretty.

Cities are dysfunctional in part because we have built them in ways that pollute our ecosphere, something that harms our health in a direct way. Ecological stupidity is also economic stupidity, and North American urban development is incomprehensibly expensive. But cities also don't work socially: their design discourages casual public contact, which is the source of strong local communities and of self-confident collective action, as Jane Jacobs pointed out over forty years ago.

Arguing that feelings of separateness from nature are mirrored in lack of connection to authentic urban life, Fowler points to numerous examples of humans who have transcended this culture of separation. In many places, peoples' energy and intelligence have produced, over time, an intricate web of land uses and interpersonal relationships that are subversive of the dominant culture of corporate urban development. At the same time, the social and economic vitality of these places shows an ecological intelligence that cannot be planned from above.

Edmund P. Fowler taught local government and environmental politics at Glendon College, York University for many years. His books include *Rites of Way: The Politics of Transportation Policy in Boston and the U.S. City* (1971), *Building Cities That Work* (1992) and *Urban Policy Issues: Canadian Perspectives* (2002).

Printed in
August 2004
at Gauvin Press Ltd., Gatineau, Québec